Basic Carpentry Illustrated

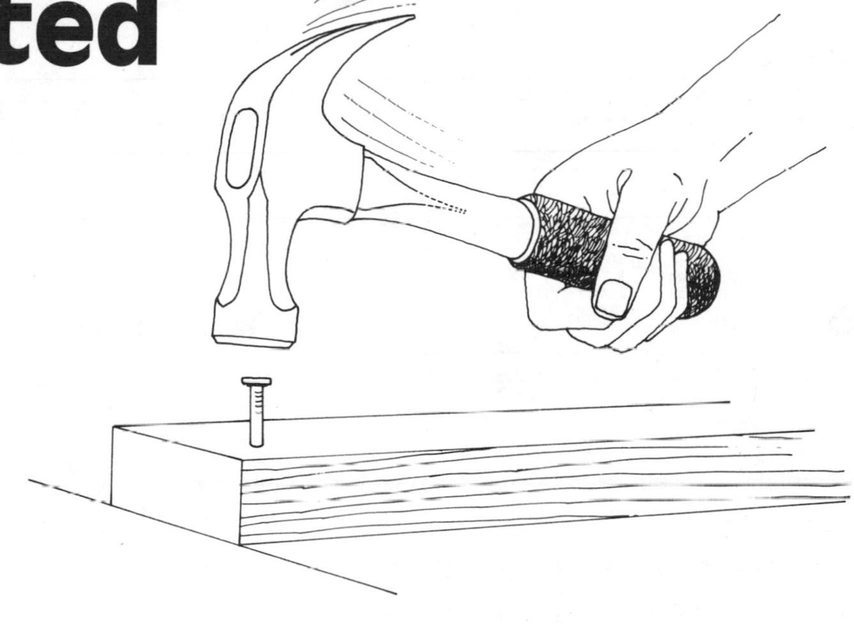

By the Editors of *Sunset Books* and *Sunset Magazine*

LANE PUBLISHING CO. • MENLO PARK, CALIFORNIA

Edited by **Donald W. Vandervort**

Design and Layout: Lawrence A. Laukhuf, John Flack

Illustrations: Joe Seney, Ted Martine

Cover: Photograph by George Selland, Moss Photography

Design Consultant: John Flack

Editor, Sunset Books: David E. Clark

Fifteenth printing June 1982

 Library of Congress No. 72-77140. ISBN 0-376-01014-2. Lithographed in the United States.

CONTENTS

Planning...

the Key to Good Carpentry

Let's assume that you've never before considered yourself a carpenter—not even a *basic* carpenter. How, then, can you recognize the symptoms that tell you it's time to become one? Perhaps your bunk-bedded children are pressing you for private rooms. That's a symptom. Or you think that your life style would improve if your garage or basement were converted into a recreation or pool room: another symptom. Even if your symptom is only general—a vague feeling that you'd like to revive the house in some small way—the time may have arrived for you to equip yourself with basic carpentry knowledge.

This book is designed to show you how to successfully complete carpentry jobs ranging from paneling a wall to installing a sliding glass door. Here you'll find information on tools, materials, and techniques for many projects.

Simple projects, such as building the bookshelves on page 57 or paneling a wall (page 51), are good starting points for beginners. They can help the inexperienced carpenter feel at ease with tools and become knowledgeable about materials. More complex projects, such as installing a sliding glass door, call for more experience.

Yet even extensive projects usually break down into a series of simple tasks. Adding a sliding glass door requires the ability to cut through small framing members, to measure, and to do light fastening work. All of these tasks are within the range of any home carpenter. When combined, though, they require the ability to think through a tightly organized sequence of steps. That planning procedure —the most difficult part—is the key to successful carpentry.

BASIC STEPS IN PLANNING

A carpenter must carefully think out how he is going to move through a job, from planning necessary materials to thinking through the project steps.

Rely on Checkpoints

One of the most efficient methods for keeping track of all the planning steps in your project is to set up a good system of sequential checkpoints. You can avoid forgetting important details by recording everything on a list.

Generally speaking, your outline of checkpoints should first include a rough look at each step of the job at hand. Your planning procedures will then become more and more refined as you get into the actual work.

Taking a survey. Surveying the situation is an important first procedure. If you plan on remodeling your basement, go down and examine it. Figure out how you want it to look, taking into account your family's needs and estimates of the cost and skills involved. If you think it might help in the next planning phases, draw up a floor plan or a rough sketch of the basement, including changes you plan.

Choosing building materials. Deciding on the materials to use is the next step. What materials will give your project the appearance and the degree of durability required? How do varying materials compare in regard to their cost and ease of application?

Many materials may be involved. Be sure not to overlook any of them. If you're remodeling your basement, will you want a suspended ceiling? What about wall paneling: would you prefer solid boards or a sheet paneling, such as plywood or hardboard? Do you plan on covering the concrete floor? If so, will you lay resilient floor tiles or add a wooden floor? What about new windows or doors? Find answers to questions about all materials involved in the project before you begin work.

You can locate information on such general materials as lumber and nails in the "Materials" chapter beginning on page 24. If you are interested in materials used in certain parts of the house (such as paneling or flooring), turn to those chapters and their subsections that specifically locate your project.

Locating assistance. The third step—finding help—requires a bit of reading and possibly some telephone calling. After locating the information on your project, read it carefully. If you feel that part or all of the project is outside your abilities, consider hiring professional help—a general contractor or a carpenter. If you wish, you can probably work out an agreement allowing you to work along with your hired help. When inquiring about professional help, it is usually wise to get more than one price estimate.

TOP PRIORITY: NEW ROOF

- [x] 1. Estimate difficulty, time involved.
- [x] 2. Select time to begin.
- [x] 3. Choose types of materials.
- [x] 4. Measure roof, figure pitch.
- [x] 5. Check codes, regulations.
- [ ] 6. Hire helper.
- [ ] 7. Get rough cost estimate.
- [ ] 8. Secure financing.
- [ ] 9. Gather tools, equipment.
- [ ] 10. Order materials.
- [ ] 11. Prethink construction procedures.
- [ ] 12. Complete work.
- [ ] 13. Clean up.

A CHECKLIST can help you approach procedures in an organized way.

Checking building codes. Be sure to check building codes and get a building permit when you need one. If you're in doubt about this, your local building inspector can tell you whether or not you'll need a permit for a particular project. Local building codes set standards for most large-scale construction projects. At times, the regulations seem a bit rigid, but they'll insure that your house will be sound and safe.

Deciding on tools. Open your tool box and ask yourself: how complete is it? Do you have the variety of tools required for all of the tasks involved? Many jobs require one or more special tools too expensive to buy for one use. In this case, rental agencies can be a real aid; most have everything from cold chisels to jack hammers.

Refine Your Plans

Up to this point, your planning has primarily involved making rough estimates. Now it's time to begin refining your planning.

Measuring and ordering materials. With the writing tablet, pencil, and a measuring tape in hand, head for the project site and make the measurements you'll need for ordering materials. Compile a list of all necessary items: their quantities, their sizes, their grade, and so forth. For a big job, telephone different supply stores to get a sampling of price estimates. For figuring most construction materials, you should add a margin for mistakes.

Planning sequence of steps. Do you remember hearing of the man who built a boat in his basement and was unable to get it out of the house? His experience shows that it pays to plan out *all* working steps carefully. Think through the sequence of all procedures you plan to follow in order to avoid running into perplexing problems. For example, if you are adding a large window in a wall, you must think of temporarily bracing the ceiling before you begin cutting into the wall. If you overlook this step, your ceiling may end up on the floor.

Allowing enough time. Plan your day's activities so you won't get caught in the middle of an ongoing project. Ask yourself beforehand: how much time will the project take? For most projects, it is better to allow extra time than to be caught short. If you're pouring concrete for deck posts, for instance, you don't want to have to stop halfway through the job. If you do, the unused concrete will harden, going to waste.

Store Materials for Efficiency

To simplify and ease your work, plan the handiest place to store tools and materials. You'll want to keep them very close to your work but not placed so that they'll be in the way. Locate the most efficient spot for everything. For instance, if you are working with large plywood sheets, find the location requiring the least movement of the sheets from cutting to installing.

HOW TO USE THIS BOOK

This book is organized to clarify the steps in basic carpentry, making construction work easier for you. Sections on tools and materials are at the beginning. Look here to find out about the many different tools and materials available and their proper use.

If you have a specific project in mind, look up the chapter that covers your project in the table of contents. For example, if you want to apply siding, find "Walls" in the table of contents. Under "Walls," locate "How to Apply Your Own Siding," and turn to page 42.

A visual framing index and covering index, located on pages 38 and 39, can help you quickly find information for projects in different parts of the house. Another aid for quickly finding information is the index on page 88.

When working with the book, it is a good idea to read the chapter introduction, familiarizing yourself with the framing in that area, before moving on to the specific subsection you are concerned with.

If a particular project is not listed under one of these subsections, it is probably because that job is considered to be beyond the scope of basic carpentry.

Basic Tools

The sea otter has an unusual way of using a tool. Diving to the sea floor, he collects a clam and a small, hard stone. Then, bobbing along the surface on his back, he lays the clam on his chest and pounds it with the rock to break it open. Because he can't lug a large stone to the surface, and because a tiny one won't crack the shell, the otter is most particular about the stone he chooses.

In the same way, tool selectivity is the key to good carpentry. Hundreds of different types of tools are on store counters today. As a rule, the greater the variety of high-quality tools you have available for doing common carpentry projects, the easier your work will be because you'll almost always be able to select the right tool for the job.

Tool quality is important. High-quality tools almost always last longer and give better service than cheaply-made tools. Good tools are generally made from better materials and are lighter and stronger than their bargain-table brothers.

If you handle good tools with respect, maintain them well, and store them where you can find them, they should serve you well for years. By keeping them clean, properly lubricated, and in good working condition, you'll find they will repay the attention you give them many times over.

In this chapter, tool types are divided into five main categories: fastening tools, tools for drilling holes, cutting tools, tools for layout and gauging, and clamping tools.

Fastening Tools

Because the art of fastening pieces together is basic to carpentry, you should be familiar with the use of hammers, staplers, screwdrivers, wrenches, and pliers. Mastering the few simple techniques mentioned here will cut down on the number of bent nails and burred screws that you may otherwise collect. (For some, the savings in band-aids alone could be significant.)

CLAW HAMMERS

Most people are familiar with a claw hammer's appearance, but slight differences in the tool's size, shape, and weight determine the kind of hammer you should select for any specific project.

Two main types of claw hammers are the *curved* claw and the *ripping* claw. The ripping claw is fairly straight; it is chiefly designed to pull or rip pieces apart. The rounded claw of the curved claw hammer offers more leverage for nail pulling.

When buying a claw hammer, choose a hammerhead that is forged, hardened, and tempered from high-quality steel.

Notice that hammer faces are made both flat and slightly convex. The convex type allows you to drive a nail flush without marring the wood's surface. The mesh type is for rough framing work.

How long you want your hammer's handle to be depends on the kind of work you plan to do. Because long handles provide more leverage in the swing than short ones, they are generally used for framing work.

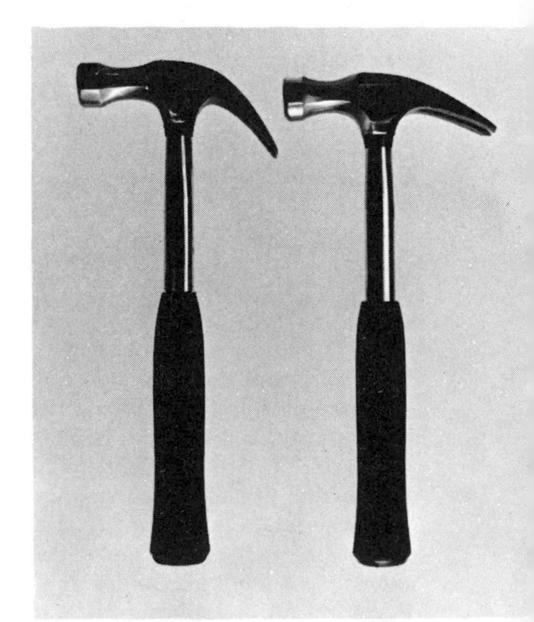

CURVED CLAW, at left, pulls nails well; at right is straight claw.

Short ones, though, are meant for finishing work. Wood handles will break before steel handles, but some carpenters prefer the feel of wood.

The head's weight is also an important consideration. Sizes range from 5 to 20 ounces. Pick a weight that is comfortable but not too light. Your arm may tire sooner swinging a light hammer for heavy work than it would wielding a heavier hammer. A 14 or 16-ounce head is popular for finishing, but a 16 to 20-ounce is more suitable for framing work.

To start a nail, hold it between your thumb and forefinger near the head and give it a few taps with the hammer. By holding it near the head, you'll just knock your fingers away rather than smash them.

Try to drive nails in at a slight angle. They hold better when driven this way than when they are driven in straight.

Once you've carefully started a nail, pound it in with a few hard strokes. Use a nailset to set finishing nails below the wood's surface.

To hammer properly, grip the handle near its end. Swing it with a full

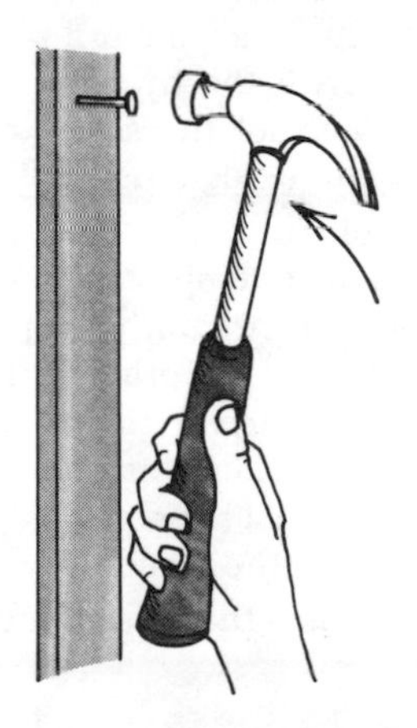

stroke and hit the nail squarely with the center of the hammer face, keeping the pivot point at the handle's end level with the nailhead.

To pull out a nail, wedge the V of the claw around the nail's shank between the nailhead and the wood, and then rock the hammer backward. To guard against marring a fine wood surface, place a putty knife or flat stick between the ham-

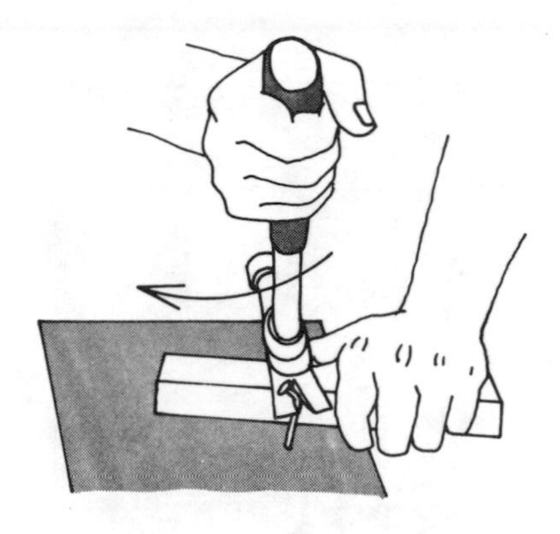

merhead and surface. For added leverage when pulling a balky or long nail, place a thicker piece of wood under the hammer and rock the hammer sideways, curling the nail over the side of a claw instead of rocking backwards. Doing this will also keep the nail hole's size small and prevent breaking the hammer's handle.

Toenailing and clinching are often necessary. When you can't nail from the flat of one surface into the end of another (as in nailing wall studs into place), you must nail at an angle through the side of one board and into the surface of the other. This is called "toenailing." Choosing the proper angle to drive the nail is important; the nail should penetrate well into the wood in both pieces.

Clinching nails takes place when you are laminating two boards together. To do this properly, hammer at least three nails through both boards until their heads are firmly seated in the top board and at least 1 inch of their points sticks through the bottom board. Turn over the two boards and keeping the nailheads seated snugly, bend over the nail points in the direction of the wood grain. Following these steps, it's a cinch to clinch.

STAPLERS AND NAILERS

To make some carpentry fastening jobs easier, power-driven staplers and nailers are available. Their source of power varies. Lightweight staplers are squeezed or slammed against the surface being fastened. One heavy-duty type is mallet driven; more sophisticated heavy-duty models are driven pneumatically or electrically.

The inexpensive lightweight types hold a moderate amount and several sizes of staples. Driving the fastener in with one squeeze or stroke, they free your other hand to hold the material. Lightweight staplers are handy for fastening building paper, insulation, or ceiling tiles into place.

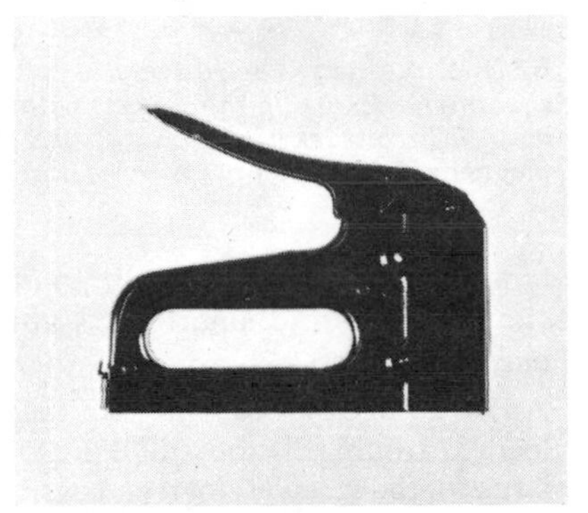

SQUEEZE-TYPE STAPLER rapid-fires for lightweight fastening jobs.

Some heavy-duty staplers and nailers may hold two or three sizes of staples or nails in quantities of 300 or more. Electric or pneumatic types are limited by their cords or hoses. Heavy-duty nailers are generally used only for assembly-line jobs, such as nailing down flooring boards or roof decking. Because they're expensive, it's probably best to rent one for specific jobs from your local tool supply.

SCREWDRIVERS

The lowly screwdriver vies with the hammer as the most frequently reached-for tool in a homeowner's collection. Its usefulness in carpentry is far greater than its simplicity might suggest.

Most people are familiar with the ordinary screwdriver, but there are many situations where an ordinary driver will not work. You need a Phillips screwdriver to handle a Phillips screw, and you'll want at least a small selection of the other

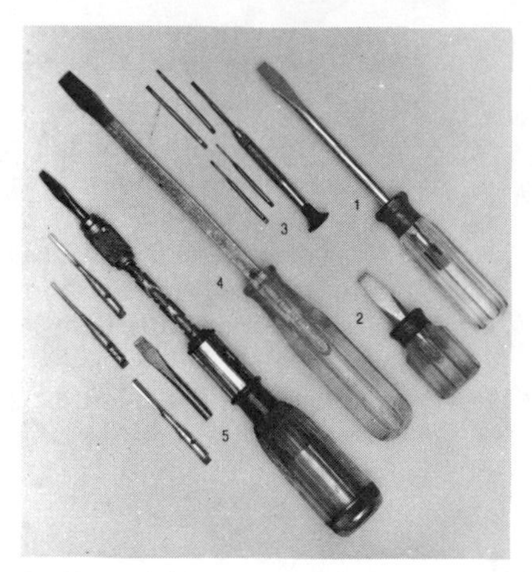

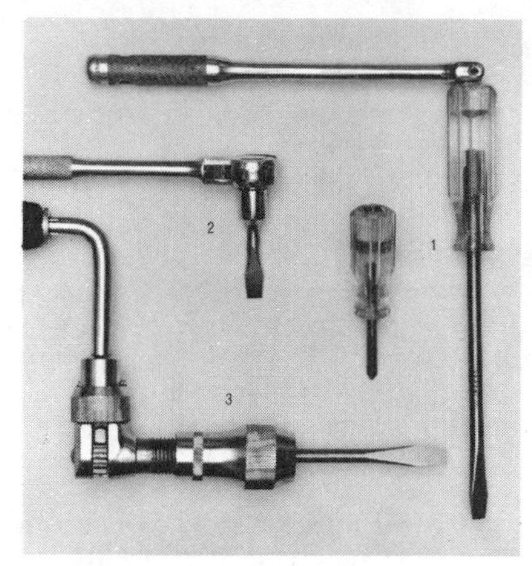

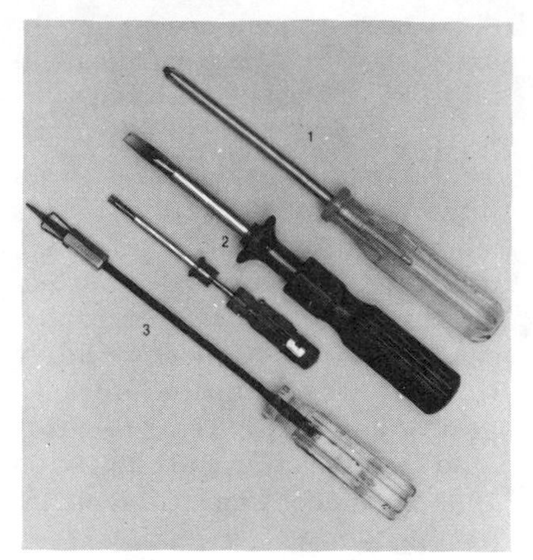

HANDY ordinary screwdrivers at left: (1) and (2) long and short-shanked; (3) jeweler's, with five tips; (4) heavy-duty, with square shank; (5) spiral-ratchet, with tips and bits. Leverage drivers in center include (1) ordinary screwdriver and Phillips, both fitting a socket wrench; (2) driver tip for a socket set; (3) driver tip for a brace. Good screw-holding drivers at right: (1) magnetized Phillips tip; (2) large and small two-piece wedge tips; (3) sliding spring clip.

types shown to handle the different-sized screws and small bolts you may encounter.

An ill-fitting screwdriver may lead to trouble. If too small or too large for the screw's slot it can easily burr the screw head or gouge the work. For this reason stock your tool box with at least three different sizes of screwdrivers.

Many jobs are bungled by using damaged screwdrivers. If they are so worn and blunted that their tempered tips are gone, or if they have been overheated, no amount of sharpening will repair them. Cheap screwdrivers, usually made of soft steel with little temper, are generally a poor investment.

Here are some tips for buying screwdrivers:

1. When buying a large, ordinary screwdriver, choose one with a square shank. Its advantage becomes clear when you have a stubborn screw to seat or loosen for you can fit a wrench on the square shank to apply extra leverage.

2. A long screwdriver lets you apply more power than a shorter one with a tip of the same size. The long one has the added advantage of being less likely to tilt in the screw slot. Save the stubby screwdriver for spots where you're cramped for working space.

3. If you are installing especially large screws, a screwdriver bit in a carpenter's brace works easily and rapidly. To drive a great many screws, a power drill equipped with a speed-reducing screwdriver attachment works even better.

4. Sometimes carpenters use spiral-ratchet "speed" drivers longer than the one shown. Although they drive screws rapidly, they slip easily unless you have had practice.

5. A magnetized screwdriver of the Phillips type holds Phillips screws well, but slotted screws sometimes slip out of an ordinary magnetized driver. Wedge-type drivers hold slotted screws well under all situations.

Hints for Using Screwdrivers

In all but the very softest wood, you should drill lead holes for screws. In hard wood, the holes should be approximately the diameter of the screw's core (the diameter of the threaded part, measured across the thread valley).

In soft wood, holes can be about two-thirds the core diameter and smaller still in the wood's end grain. When using long screws, you should also drill larger shank-sized holes in the wood to the depth of the screw's shank (the unthreaded portion). Or use a screw pilot bit that cuts both holes at once.

If a good screwdriver requires undue muscle and tends to burr the slot when the screw is only partially in, remove and discard that screw, then enlarge the lead hole for another. With brass screws (especially long ones going into hard wood), it is a good idea to drive in and remove a similar steel screw first in order to form threads for the softer screw. Rubbing a bit of soap or wax on a screw makes it easier to drive in.

To remove a stubborn screw, first heat it with the tip of a soldering iron. Then let the screw cool and remove it, using a driver.

A helpful hint: have you ever struggled to keep a regular screwdriver's tip from slipping out of a screw's slot? To prevent this, tape screw and driver together with a small piece of masking tape; remove the tape when the screw is partially driven into the wood.

Remember that your screwdrivers

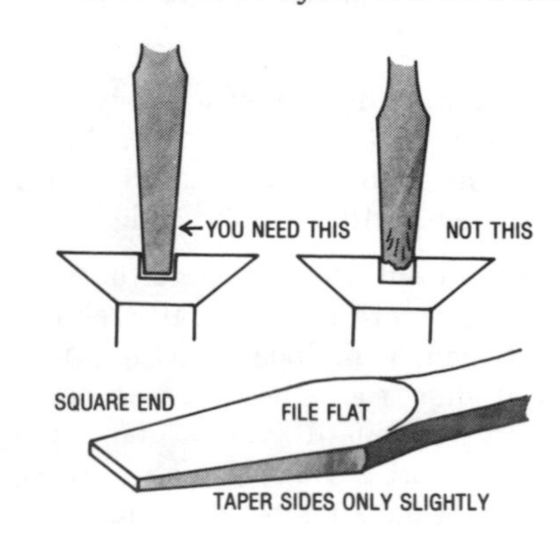

need to be kept properly shaped. An ordinary file will reshape most of them easily and smoothly. On a power grinder, use the finest wheel, being careful not to overheat the driver's tip.

WRENCHES AND PLIERS

Wrenches and pliers fall into the category of troubleshooting tools, tools that you turn to when problems arise in a carpentry project. When your hand ceases to apply enough leverage to turn a nut onto a bolt, a wrench will supply the extra muscle power you need. Choosing one that fits snugly will prevent the nut's square or hexagonal shape from becoming somewhat rounded.

Open-end wrenches are available individually or in sets, in both open-end and box-end styles (or a combination of both, as shown). Each is made to fit precisely only one particular size.

The open-end wrench is handy for rapidly turning a nut where access limits you from "spinning" it on. Using a box-end wrench, you can apply severe pressure without the wrench slipping off.

If you prefer not to buy a series of wrenches, purchase an adjustable *crescent wrench.* You can adjust its open-end jaws to fit a variety of nut sizes. When using it, periodically check to be sure jaws are tight around the nut.

Monkey and pipe wrenches should not be used for turning nuts. Because their jaws are not parallel, they will tear nuts and bolts.

A pair of pliers is also a very useful tool. With adjustable pliers you can hold a bolt's head while turning the nut with a wrench, twist wire, bend metal plates, grip hard-to-hold objects, and remove broken nails and screws from wood surfaces.

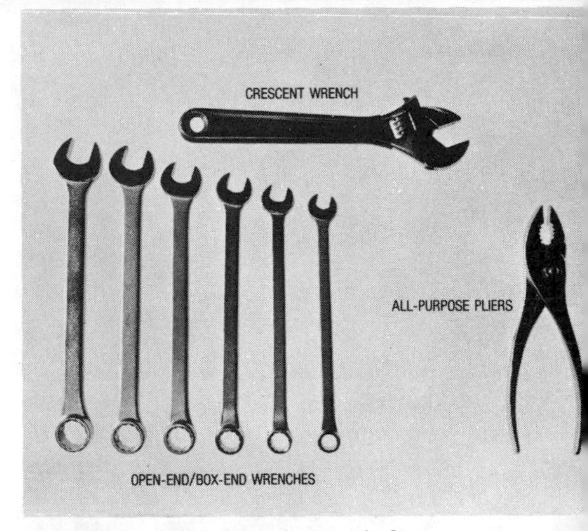

SET OF WRENCHES and pair of pliers are useful for fastening jobs.

Cutting Tools

". . . the most unkindest cut of all." True, this was spoken by Shakespeare's Mark Antony in his funeral oration for Julius Caesar, but it's also been voiced by many a carpenter who has spoiled a piece of lumber by sawing or shaving off too much. To keep you from falling prey to such a cruel slice of fate, this chapter offers information on choosing and using the three main types of cutting tools: saws, chisels, and planes.

SAWS

Saws, like most carpentry tools, differ according to the purpose for which they're made. The two main types are handsaws and power saws. The primary handsaws used in carpentry are the crosscut saw, ripsaw, backsaw, compass saw, keyhole saw, and coping saw. Although the category of power saws includes the jigsaw, bandsaw, radial-arm saw, and table saw, this book focuses on the two used for on-the-spot carpentry: the portable circular saw and the saber saw.

Handsaws

Shape of the saw, blade size, and position and number of teeth along a handsaw's blade determine the type of cutting it will do.

The term "point" applied to a saw indicates the tooth size and number of teeth per inch. Thus an 8-point saw has 7 teeth per inch. A 12-point saw has 11 smaller teeth per inch, and cuts more slowly and smoothly.

To prevent a saw from binding in the cut, teeth are "set" (bent outward to produce a cut wider than the blade). A crosscut saw has alternate teeth bent outward (about ¼ of the blade thickness) in opposite directions. So does a ripsaw, but the set is ⅓ of the blade thickness.

Crosscut saw enjoys the widest use. Made to cut across wood grain, it also does all-purpose cutting of plywood, hardboards, and other wood-base materials.

Saw lengths vary from 12 to 26 inches. Longer lengths are easier to use, but a short saw becomes indispensable when you are cutting in tight areas.

Fine-quality saw teeth have a cutting edge on both sides. Because of the set, they actually cut two grooves, chipping out the excess wood between.

For easiest cutting, saw at a 45° angle to the surface. If you are right-

CUT AT 45° angle with crosscut saw; cut at 60° with ripsaw.

BACKSAW is for fine cutting jobs.

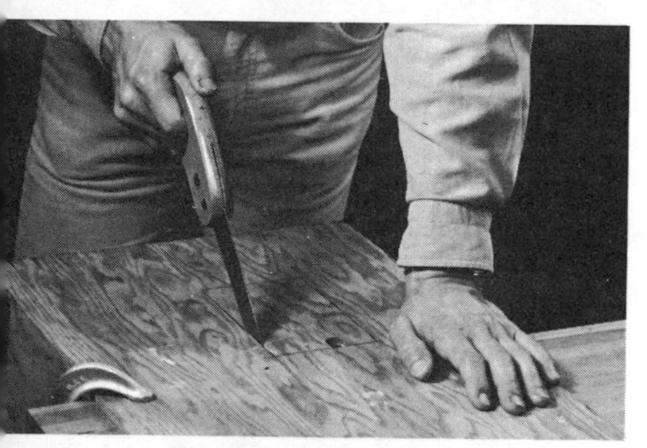

COMPASS SAW makes cut-outs.

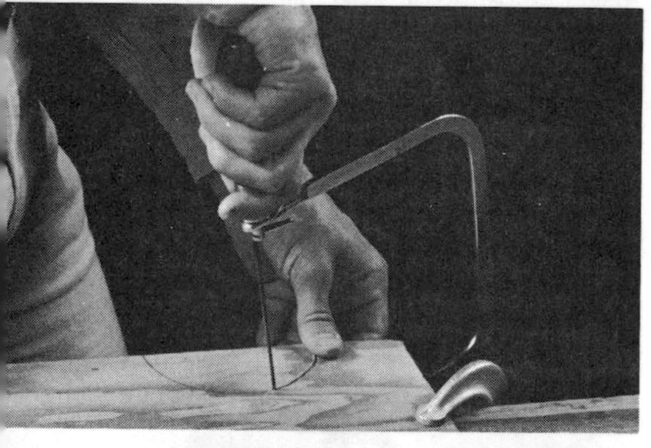

COPING SAW is for irregular cuts.

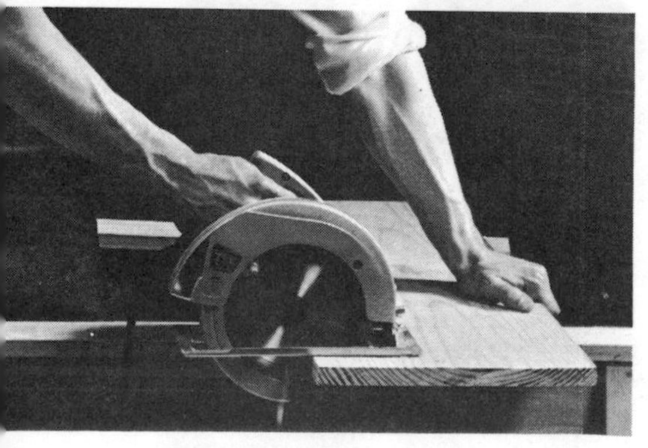

CIRCULAR SAW makes cutting easy.

handed, watch the cutting line from the left side of the blade. If you are left-handed, watch it from the right side. Remembering that teeth cut on the forward stroke, start the cut with short strokes, progressing to smooth, full strokes. Prevent the wood's underside from breaking out as you cut by first scoring it with a knife. As you near the end of the cut, support the excess piece so that it won't break off.

If the blade veers from your cutting line, twist the handle *slightly* to the opposite side until the blade returns. Those who have a persistent problem in keeping the cut straight should clamp a straight board along the cutting line for the side of the saw to ride against.

Ripsaw has small, chisel-like teeth that cut rapidly in line with the wood grain. It generally has only four to six teeth per blade inch. Cutting on the forward stroke, it is best held at a 60° angle.

Backsaw, with a thin blade having 11 to 16 teeth per blade inch, is primarily used for finishing work in which precise cutting of straight lines is important. Rectangular in shape, the backsaw derives its name from a metal reinforcing strip that runs the spine's length to keep it from bowing.

Unlike the blades of the crosscut and ripsaw, the backsaw's blade is held parallel to the cutting surface. A *miter box* is often used to support the material securely and precisely guide the saw.

A compass saw makes cut-outs and curved cuts. Its 12-inch-long blade is pointed at the top and less than an inch wide at the butt.

When cutting curves with a compass saw, saw perpendicular to the surface. To begin a cut-out, insert the blade in a previously-drilled hole. Once the cut is started, use a regular handsaw to finish a long, straight cut.

The coping saw has a thin, wiry blade held taught in a small rectangular frame. It makes very thin, accurate cuts and saws along very tight curves with ease, but cutting is limited to surfaces that the frame will fit around. Both wood and metal-cutting blades will fit in the frame. Teeth point toward the handle, cutting on the pull stroke.

Portable Circular Saw

Your portable circular saw pays for itself many times over as soon as you begin to make relatively straight cuts in unwieldy pieces of lumber or paneling. Using this saw, you can cut painlessly ten times the amount of wood that you could cut with a handsaw. And, unlike a table saw, which requires you to round up a couple of helpers to carry and maneuver the wood, the portable circular saw can be taken to the project and operated by you alone.

Choose a saw that is large enough to handle the material you'll be working with. A 6-inch blade will cut about 2 inches vertically, 1½ inches at a 45° angle, and rip 1-inch-thick lumber. A 7-inch model is better for extensive cutting.

Circular saw blades. While the blade that comes with a portable saw is suitable for most woodcutting, a number of specialized blades can refine and extend the saw's usefulness. Armed with a little knowledge, you can choose the proper one for a difficult job and avoid the expense of a sophisticated blade you don't need.

A blade's appearance indicates its performance. A few large, wide teeth will give a fast, deep, rough cut, whereas many small, narrow teeth will cut more slowly but cleanly. Most blades fall between these extremes, sacrificing some speed and durability for a measure of precision.

Listed below are the more common types. Though the dado blade shown on page 11 won't fit a portable circular saw, mounted on a radial-arm or table saw, it simplifies cutting grooves. You might consider two blade refinements: nonstick blades, at an average 30 percent added cost, resist friction and rust; and, coarse-cutting carbide-tipped blades ($9 to $13) have exception-

VARIETY of blade edges is shown at left: (1) toothless metal cutter, (2) combination blade, (3) plywood blade, (4) rip blade. At right, dado set (front) cuts various grooves. Masonry blade (rear) cuts stone.

ally long lives. (Use carbide blades carefully; they can chip irreparably if they hit a nail.) Make sure the blades you buy will fit your saw's arbor (axle). Here are a number of available blades that you may use:

Rip blades have wide, chisel-like teeth for with-grain board ripping. The wide teeth may slow the blade in heavier stock. Don't use for cross-grain cuts, they'll splinter the wood.

Cutoff blades have much finer teeth for squaring and clean cross-grain cutting. Ripping wood will bog them down.

Plywood blades are finer still and especially tempered to resist the abrasion of plywood glue lines and to avoid splintering.

Combination blades combine rip and crosscut principles into a cutting edge suitable for everyday needs, sacrificing only a part of the speed, stamina, or precision of a specialized blade. The many types sold can be judged by tooth size with one deceptive exception: some models have groupings of fine teeth separated by deep recesses called "gullets." These gullets remove sawdust from the cut, allowing stress-reducing metal expansion. Don't confuse them with large teeth.

The finest-cutting combination blades are called "planers" because they leave very smooth edges. Although they dull quickly in heavy stock, they won't, like most combination blades, raise splinters when crosscutting plywood. Hollow-ground combinations with bodies narrower than the teeth give a particularly clean cut without sticking.

Flooring blades are specially tempered for cutting old lumber containing occasional nails or wire.

Cutoff wheels are abrasive blades designed to cut a variety of metal and masonry surfaces. They greatly speed the cutting of difficult surfaces, but they wear out quickly and must be used in strict accordance with enclosed safety instructions.

Operating a portable circular saw. The portable circular saw gives the carpenter the advantages of flexibility and convenience. But in careless hands it is also one of the most dangerous of the power tools.

These recommendations for safe use, given by the *Power Tool Institute*, can be applied to the use of almost every power tool:

1) *Keep Environment Safe.* Always work in a dry, well lit area with plenty of room for maneuvering. Your work area should be neat and equipped with the necessary clamps for securing the job. A power saw should be operated without interruptions or distraction. All visitors, especially children, should be kept from entering the work area while a power saw is in operation.

2) *Know Your Tool.* Be sure to read your owner's manual. A first hand acquaintance with your power tool is essential. Know its capabilities and limitations. Choosing right tool for the job is essential.

A power tool should be properly grounded unless it is double insulated (see page 19). Disconnect saws before changing blades.

Saw blade guards should never be removed or blocked. Manufacturers have designed these guards to allow for easy operation of the power saw.

3) *Dress Safely.* Be sure you're wearing proper apparel before starting the job. That means no loose fitting clothing or jewelry. Safety glasses only take a minute to put on and can be worth their weight in gold.

4) *Operate Safely.* While operating the saw, avoid awkward positions and never stand in line with the blade. Always let the saw do the work. Use a slow, steady motion when pushing the saw through the work. Special care should be taken when cutting through material with knots. The feed should be slower, and the operator must be alert for binding and a quick stop.

Before operating the power saw, examine the material to be cut for any objects (such as nails) that could dull or bind the cutting blade. Pick the right blade for the job. When using accessories, be certain that they were designed for your saw and for the job at hand.

After completing the work and turning off the tool, remember to disconnect the power plug and remove the switch key. Store it out of the reach of children.

5) *Observe Proper Maintenance.* Keep the tool in the best operating condition. Always be sure that it's sharp, clean, and lubricated according to specifications. Remember that a sharp blade not only cuts cleaner but also is less likely to bind. Be absolutely certain to disconnect tool before servicing.

SABER SAW cuts partially into plate.

Power Saber Saw

Versatility is the saber saw's claim to fame. It can do the work of six different saws—crosscut saw, rip saw, band saw, keyhole saw, hack saw, and jigsaw.

This portable power tool has a high-speed reciprocating motor that drives any of several interchangeable blades to cut wood up to 2 inches thick, sheet metal, plastic, rubber, leather, and even electrical conduit. In addition, you can make cut-outs in wood, paneling materials, wallboard, and plastic laminates without drilling a starting hole; cut intricately-curved lines; and cut straight lines, circles, and bevels using a guide.

Types of blades. The type of cutting a particular blade will do is determined by its teeth. All blades cut upward. Fine teeth cut smoothly but slowly; coarser teeth do faster work but often leave too rough a cut for cabinetwork. For general woodcutting, compromise by choosing a combination blade that will both rip and crosscut.

Metal-cutting blades are selected according to the metal being cut. Use a fine blade for thin sheet metal and slightly coarser blades for thicker metals. When cutting sheet metal, clamp it with a wooden batten or between two layers of thin plywood to keep it from vibrating. Keep your saw level when cutting tubular or irregularly-shaped metal by placing blocks on each side.

Using the saber saw. To avoid breaking blades, spoiling materials, and risking injury, follow the same rules listed under the power circular saw.

The upward-cutting blade may cause the material's upper surface to fray around the cut. To avoid this, scribe along the top of the cutting line with an awl, or tape the line with transparent tape.

To make a cut-out in the center of a panel, tip the saw forward onto the tip of its shoe in such a way that the blade does not contact the surface. Turn it on and gently tip it back until the blade tip cuts into the wood. Once the blade cuts through, seat the saw flat on the work and finish the cut.

CHISELS

Chisels are used to make small grooves, notches, and cuts. The type shown, with metal-capped plastic handles, can be driven with a hammer. Other types, with uncapped plastic or wooden handles, should only be driven by hand or hammered with a wooden mallet.

The set shown is a good combination to have: blade widths are ¼, ½, ¾, and 1-inch.

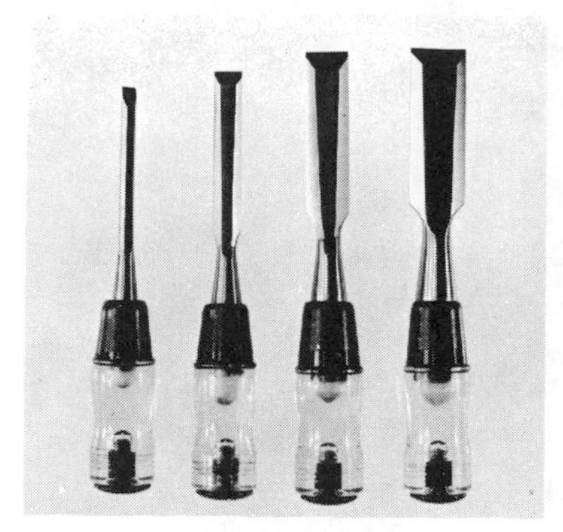

SET OF 4 CHISELS, basic for tool box.

How to Use a Chisel

Two types of cuts are made with a chisel: slicing cuts, made with the chisel alone, and mallet or hammer-driven cuts.

For best control when driving a chisel, use light taps and remove wood in small stages. Try to start a cut on the waste side of a cutting line, working back toward the line. Always make defining cuts across the wood grain before driving the chisel parallel to the grain; otherwise, the wood may split beyond your outline.

Chisels are most commonly used for making mortise cuts. Cutting a shallow mortise is easy: 1) using a pencil, begin by outlining the area to be removed; 2) score with a knife along the lines that run parallel to the grain; 3) place the chisel edge (with the bevel facing the waste wood) just inside the cutting line, lightly rap it several times with the mallet or hammer, and do the same at the other end of the mortise; 4) starting at that cut, proceed to make similar cuts about every ¼-inch, across the waste wood, moving with the blade's bevel forward; 5) slice along the lines that are parallel to the grain, hand-holding the chisel. You can then easily remove the loose center sections, holding the blade's bevel parallel to the surface.

For deeper mortises, first bore out the waste wood using a drill (see page 14).

Sharpening Chisels

Keep your chisel edges sharp at all times by whetting them on an oilstone. Though a chisel's blades are usually ground to a 25° bevel, whet it at a 35° angle (by pushing it in a circular motion across the oilstone.) Once the edge is sharp, turn the blade over, lay it flat on the stone, and slide it back and forth to remove excess metal particles.

Nicked blades should be reground to a 25° bevel. To do this, use either a fine mill file or a wheel grinder. If you use a mill file, file diagonally. With a wheel grinder, slowly move the blade from side to side across the wheel's surface, holding it at a 25° angle. To keep the blade from overheating, cool it frequently in water. After grinding the blade, whet it to a fine edge as explained above.

PLANES

Slicing off unwanted portions of wood is the job of the plane. It does the same work as a chisel. But in the case of the plane, the depth and width of its cuts are controlled by the body of the tool.

Different types of planes are designed for different jobs. Those most commonly used fall into two categories: *bench* and *block* planes.

A block plane, small enough to be held in one hand, is suitable for planing rounded surfaces and for planing end grain, or "blocking in." Bench planes are used for smoothing with the grain along a board's length.

Other planes, such as rabbeting, routing, grooving, and chamfering planes, are available for special jobs, but they are not discussed here because they're more often used in cabinetmaking than in general carpentry work.

Bench planes come in three main types, the chief difference between them being in size: the *fore* (or *jointer*) plane, the *jack* plane, and the *smooth* plane. The shorter a plane is, the more it has a tendency to ride up and down the irregularities of a surface. For this reason, a long plane is used for truing long surfaces, a short plane for shorter or more irregular work.

The fore plane is largest, measuring from 18 to 30 inches long and about 2½ inches wide. It will glide along a board's surface and shave off the high spots until the entire length of the surface is flat.

One size shorter, the jack plane is from 12 to 15 inches long. It is used for rough work, such as removing irregularities on a board's surface or edge, sometimes leaving a rough cut.

Smallest of the bench planes is the smooth plane, ranging in size from 5½ to 10 inches long. Use this plane for smoothing surfaces after using the jack plane. The smooth plane's short length limits its ability to true long surfaces.

Using a Plane

Before beginning a cut, check the angle and exposure of the iron's (blade's) edge by sighting down the plane's underside as shown. The edge should protrude only slightly through the opening and should be square. If it isn't square, adjust it by pushing the lateral adjustment lever toward the side of the iron that is extended farthest, if it is a bench plane. In the case of a block plane, release the cam cap and adjust it by hand. The plane iron of a bench plane should be mounted with bevel down; a block plane's blade should be bevel up.

HOLD PLANE upside down to sight blade setting. Cut in shearing motion.

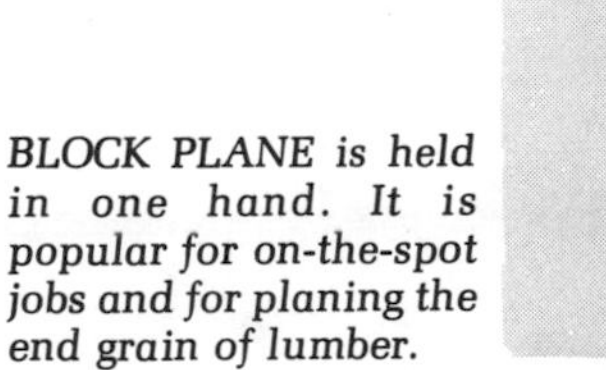

BLOCK PLANE is held in one hand. It is popular for on-the-spot jobs and for planing the end grain of lumber.

JACK PLANE is the middle size of the bench planes. A two-handed plane, it serves well for all-purpose work. Use it only in line with grain.

Slightly angle the bench plane so that it makes a shearing cut (see sketch), always cutting in the same direction as the grain. Keep cuts shallow and even, applying most force on the front knob at the beginning of the stroke, then evening out pressure at the middle, and finally applying force to the handle at the stroke's end.

Holding a block plane in one hand, apply pressure with the forefinger to the front knob if necessary. Use short, shearing strokes to cut the end grain. To prevent splitting a board's edge when planing end grain, plane inward, slightly bevel the edge first, or clamp a piece of scrap wood to the edge.

Keep the Iron Sharp

A plane's iron should be sharp at all times. Procedures used to sharpen one are very similar to those used for sharpening a chisel (see "Chisels"). The main difference appears in the edge's shape: a block or fore plane has a squared-off edge like a chisel's, but the smooth and jack planes' blades have slightly rounded edges.

Tools for Drilling Holes

Chances are that grabbing a necktie at random from the closet on an especially hurried morning isn't going to influence the quality of your work one way or the other. But in the workshop, the story is different. The particular tool you lift from the workbench to drill a hole with can make the difference between good and bad craftsmanship.

If you don't know each of the different tools' capabilities, you can spoil perfectly good tools, squander time and materials, and end up buying more tools and attachments than you need. On the other hand, you can save time, tools, and attachments if you understand the specific uses of each piece of equipment.

We generally speak of *bits* for wood (boring) and *drills* for metal (drilling). However, the term *drilling* can apply to both processes. And you'll find that a number of drills designed primarily for metal are also suitable for wood boring. Inquire before you buy if you want drills that can be used for both.

Three main types of tools are used to drive bits and drills: braces, hand drills, and power drills. Each is used for a slightly different purpose, so each is combined with slightly different bits or drills.

THE BRACE AND BIT

When it comes to boring large holes in wood, the brace and auger bit are of first importance. Using a brace to drive an auger bit, you can bore holes in wood from ¼-inch up to about 2 inches in diameter for anything from bolt or screw pilot holes to holes for dowels and pipes.

The Brace

The brace is a tool that is used much like a crank, with an attached bit. Its sweep may vary in width from 6 to 14 inches; the average brace has a 10-inch sweep. Some braces have a ratchet, a gear-like device that permits you to bore holes in tight places (such as close to walls or ceilings) without making full sweeps of the brace's handle.

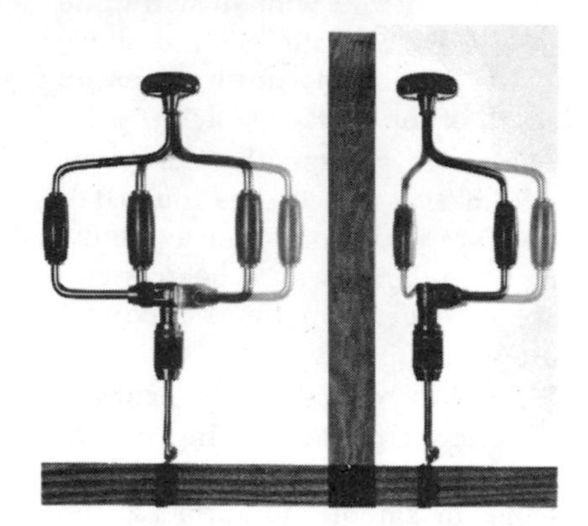

BRACE AT LEFT is used with a full sweep. At right, ratchet lets you bore hole in tight spot.

A bit is easily "chucked" into a brace. Open the jaws of the chuck only far enough to admit the bit and insert the bit as far as it will go. Then when you tighten the chuck, the bit should be straight and well-centered.

To bore a hole, center the bit's point on your mark, hold the round butt knob with one hand, and turn the offset handle with the other. Once the bit has almost penetrated the piece, either block the piece from behind or reverse the piece and drill from the other side so that the wood won't splinter (see photo).

SPLINTERING WOOD (right) occurs when you bore all the way through. Instead, finish from other side, or back hole with a block.

Bits For Braces

Though the auger bit is a simple tool, it must be properly balanced to be effective. The screw and cutters of this bit should pull through the wood at exactly the same pitch; other parts work to counteract the resistance as the spurs dig forcefully into the wood.

The size and shape of an auger bit will vary according to special uses as follows:

Length—three main sizes are specified:

Short dowel—about 5 inches.

Regular—7 to 10 inches, for all-around use.

Long ship—more than 10 inches, for boring through heavy timbers.

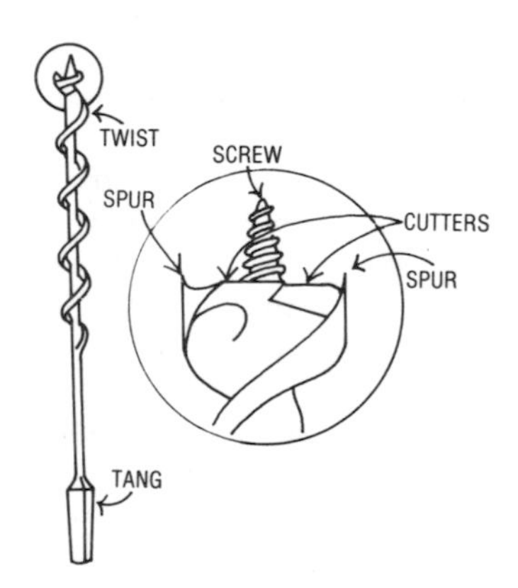

WORKING PARTS of common auger bit.

Twist type—either single or double twist, or solid center:

Single twist (hollow spiral)—throws cuttings to the center hole and easily delivers them to surface. Good for deep boring.

Double twist—accurate, less desirable for long drills since the cuttings are crowded to the walls of the hole where they may get jammed.

Solid center (straight core, single twist)—preferred when drilling hard woods.

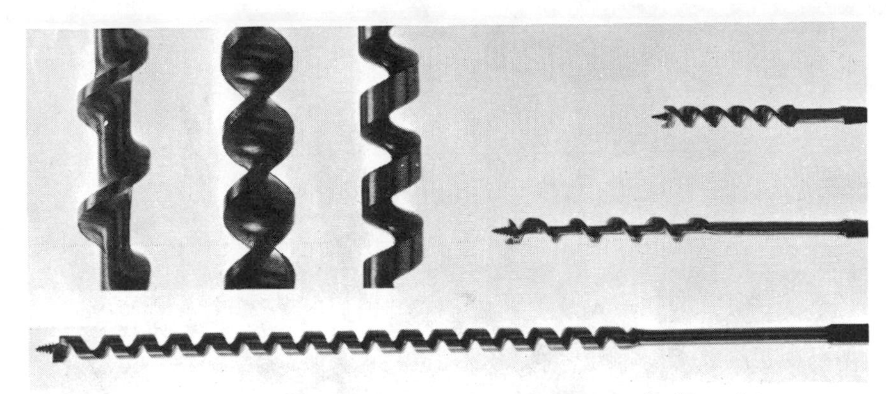

AUGER TWIST TYPES (from left): single twist, solid center; double twist; single twist, hollow center. Lengths (from top): dowel, regular, ship.

Type of bit heads—variations of spurs and cutters:

Double cutter with double spur—for general use in making smooth holes. Bores and cleans easily.

Single cutter with spur— recommended for boring in wet, green, or knotty wood.

Single cutter, no spur—for deep boring in woods with heavy grain. Makes a rough hole.

Barefoot, singlecutter, no screw—for deep boring in wet, pitchy woods or when straight boring is not especially necessary. Since it has no screw, it tends to drift with the wood grain.

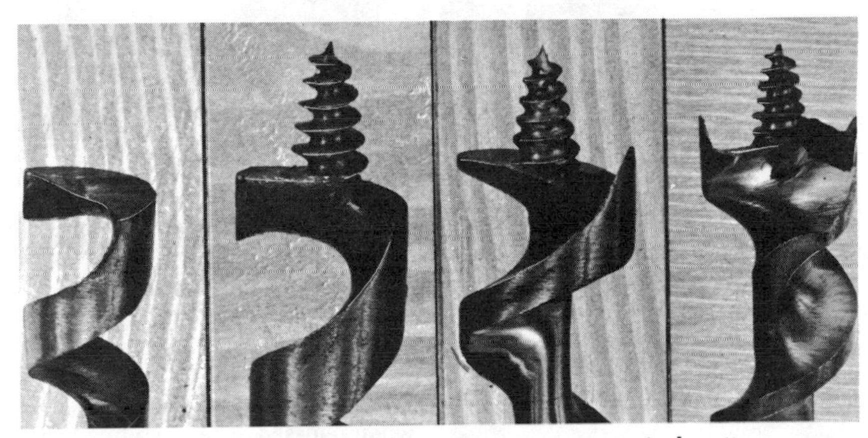

AUGER BIT HEADS (from left): single cutter, no screw or spur; single cutter, no spur; single cutter, single spur; double cutters and spurs.

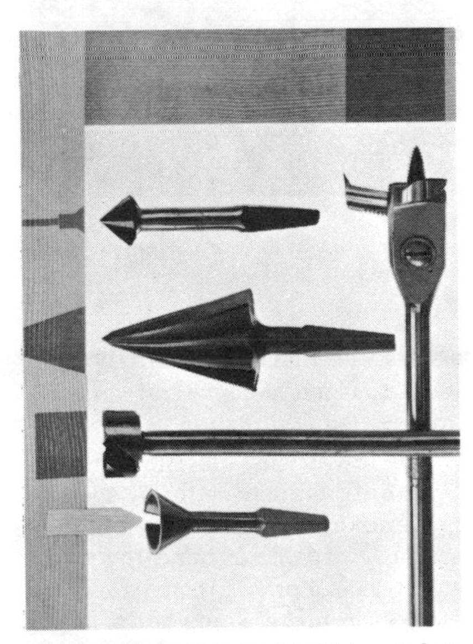

BIT BRACE ATTACHMENTS (top to bottom): countersink, reamer, Foerstner bit, spoke pointer or dowel sharpener, and (at right) expansive bit (after adjusting, make test hole before final boring).

Type of screw—its threading:

Single thread – pulls fast.

Double thread – pulls more slowly.

Coarse thread – as few as six turns per inch.

Fine thread – as many as 18 turns per inch.

For general carpentry, the best bit is probably regular length, solid center and single twist, double spur and cutter, a single coarse thread.

Other brace attachments can be used to do jobs beyond the range of the auger bit. Expansive bits are available for boring holes of varying diameters from about 1 to 3 inches. The diameter of any one of these bits will have a range of about 1 inch—say from 1½ to 2½ inches.

The Foerstner bit, having a circular rim that projects beyond the cutters, is extremely accurate but somewhat slow.

Other attachments include screwdriver bits, countersinks, dowel sharpeners, reamers, bit extensions, and bit depth gauges.

Twist bits (see "Electric Drills") can also be used in braces and are less likely to split wood.

Taking Care of Bits

If you keep bits in a soft wood box or folding cloth holder, they will need little sharpening. After using bits, turn them through a fold of an oily cloth. To protect the screw and spurs, put a dowel or cork on the end of the bit.

Nails, concrete, and metal can seriously damage the screw and the spurs of the bit. Avoid them when boring.

Sharpening an auger bit takes considerable skill and is usually a job for an expert.

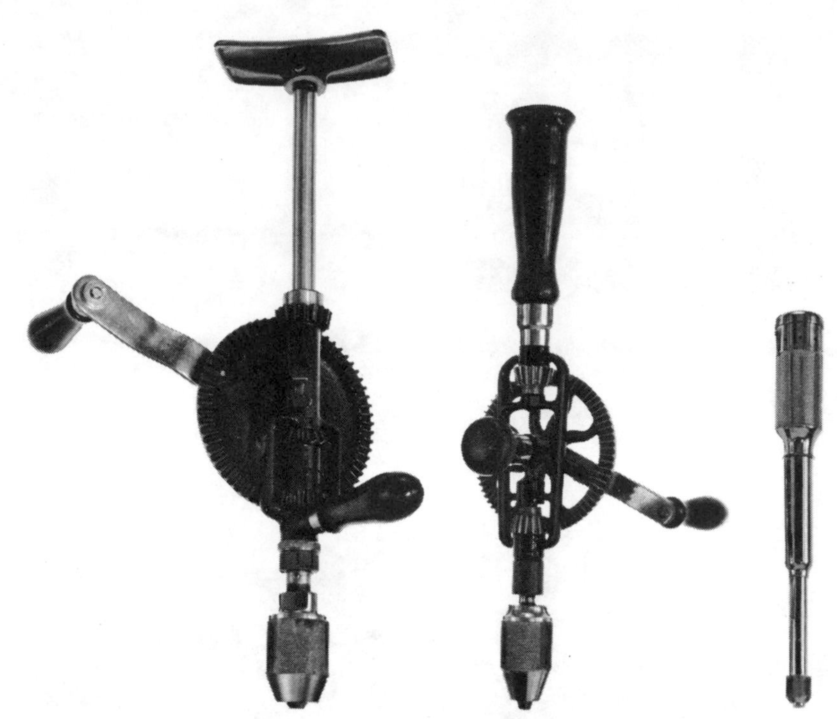

LEFT TO RIGHT: breast drill, hand drill, and automatic push drill. Latter type may have pistol grip. Some have a chuck that won't let bit slip.

HAND DRILLS

For carpentry that calls only for the drilling of small holes, use a smaller hand drill, breast drill, or push drill. The hand drill and breast drill are both "egg beater" types. The main difference between them is that the breast drill has a cross bar to lean your chest against for more pressure; the hand drill does not.

Both can be purchased with two gear speeds, and both often have a cavity inside the handle where you can keep a set of drills. Both use drills up to about ¼-inch in diameter and will make holes much more rapidly than a brace. Used for drilling both wood and metal, jaws usually take round shank drills, but some have universal jaws which also take square shanks.

Make a starting hole in wood using a brad, and center-punch metal before drilling a hole with these drills.

The push drill is used for making very small holes rapidly. A strong spring and spiral mechanism makes the chuck revolve clockwise when pushed down, counter-clockwise when released. The push bits used with it are specially ground with points that cut when rotated in either direction.

Tips for Using Hand Drills

Never put too much pressure on any hand drills. Because the drill bits are usually small in diameter, they bend or break easily. If the bit wobbles while turning, it will make an oversized hole. When drilling through metal, relieve pressure slightly before breaking through to avoid any danger of snapping the bit's point.

When the hole is drilled to the desired depth, keep revolving the drill as you remove it. If you pull it out abruptly, you may break it.

Because these bits are often quite small, always return them to their proper place after using. Doing this saves trouble in finding the correct size next time.

PUSH DRILL punches bolt holes through a simple template here.

ELECTRIC DRILL

Do you think that simple hole-boring is the sole purpose of the electric drill? Think again. This basic tool has spawned a host of accessories that can make it the nucleus of an inexpensive but versatile home power workshop

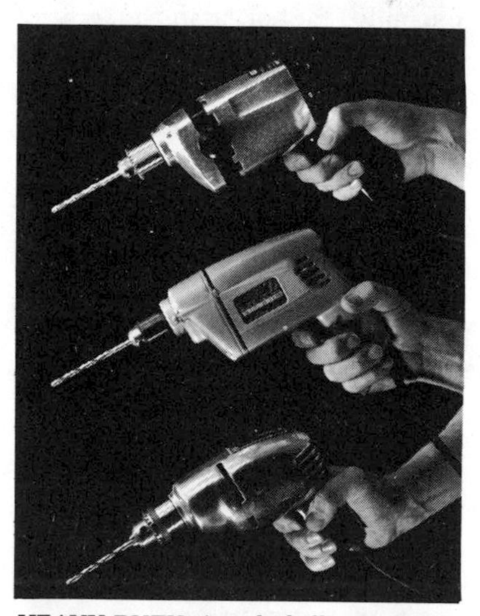

HEAVY-DUTY ¼-inch drill at top costs most; middle one has variable-speeds; bottom costs least.

The lowest cost electric drills are light-duty models with an average power of only about ⅛ hp, but they are fully capable of handling many small jobs. For frequent use, however, or to power some of the larger attachments listed below, a medium-duty or heavy-duty drill will be a wiser choice.

The most common drill is the ¼-

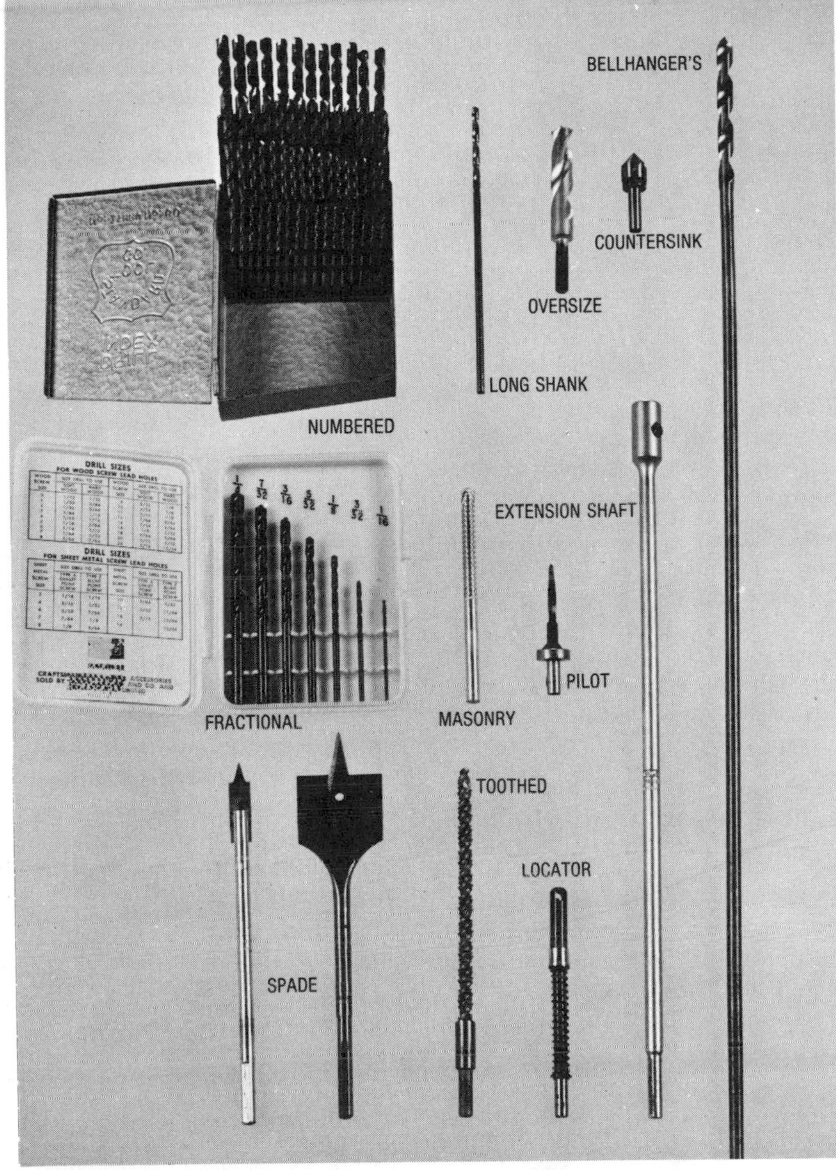

BITS for drilling small to large holes and for more specialized tasks.

inch variety. (The size refers to the diameter of the chuck, or neck, which clamps around the bit or other attachments.) Three-eighths-inch and ½-inch drills are designed more for heavy-duty use. Though they cost more, they deliver the added power necessary for difficult drilling and for the most efficient use of large accessories.

Choose a drill with a geared-key "Jacobs-type" chuck. It will center and grip a bit or attachment best.

Quarter-inch drills operate on standard household current, but all except double-insulated models require three-hole, grounded sockets to prevent shock.

Some moderately-priced drills have an electronic speed control in the trigger; the more the trigger is squeezed, the faster the bit turns. This is a helpful accessory.

Any good drill should be provided with these features: a warranty (commonly 90 days to a year), an Underwriters' Laboratories certification, a durable body, variable speed from 0 to at least 1,000 rpm, approximately 1/5 horsepower, and some type of protection against motor burnout.

Small Attachments

You can buy bits or drills for drilling metal, wood, masonry, and practically any other material with an electric drill. In addition to these, many small attachments also simply chuck into your drill.

Following is a list of some of the wide variety of bits and attachments for electric drills:

Numbered bits are small twist drills that usually come in 60 different sizes, running from a tiny number 60 (about the diameter of a straight pin) to a number 1 (slightly smaller than ¼ inch). The large selection of small sizes helps you do precise "machine-shop" work at home. A full set of 60 bits can easily cost $30 or more.

Fractional bits are the more common twist drills sized in fractions of an inch. The set shown is in "32nds" (seven bits—1/16, 3/32, 1/8, 5/32, 3/16, 7/32, ¼-inch). A larger fractional set comes in "64ths," and has 13 drills.

Choose only high-speed bits in either the numbered or fractional sizes for a power drill (low-speed bits will burn easily at ¼-inch drill speeds). Use them in wood or metal. In hard metal, use a center punch to provide a starting point or drill a small lead hole.

Long shanked twist bits, many about 6 inches long, are available in ¼-inch size and smaller.

Oversize twist bits of 5/16, 3/8, 7/16, and ½-inch sizes are made with ¼-inch shanks to fit in your drill. In hard metal, these tend to overload the drill. Make a smaller lead hole first, letting the drill cool if it heats.

The countersink shown works in both wood and metal on a ¼-inch drill. At high speeds, a countersink often tends to chatter. If it does, run your drill on "slow" (if it is multispeed), or switch it on and off to reduce speed.

Masonry bits have tungsten-carbide tips that chew slowly through concrete and stone, somewhat faster through mortar joints and brick. The larger ones require considerable power. A ½-inch bit with reduced shank is the largest for a ¼-inch drill.

Pilot bits are made in various sizes for different wood screws. One will drill the proper lead hole for a screw's threads, a larger hole for its shank, and countersink for its head—all in one operation. Handy when you are installing a quantity of screws, a pilot bit can often be used for screws a size larger or smaller than the one specified.

Spade bits are for wood and plastics. Because they are inexpensive, cut clean holes, and seldom bind, they are widely used today in ¼-inch drills. They're usually available in sizes from ⅜ to 1½ inches.

The toothed bit shown is ¼-inch size and has a normal twist-drill tip for piercing the work. It also has sharp saw teeth along its long shank that will enlarge a hole, cut slots, and even saw sideways through wood, plywood, and wallboard much like a saber saw.

A locator bit will precisely locate the holes for screws when you are installing hardware. Available in several sizes, it has a pilot-type bit inside a metal guide tube. The tube centers itself in the holes in the hardware, the aligned bit drilling on into the wood as pressure is applied.

An extension shaft is a sturdy shaft about 12 inches long in which you lock a wood bit (usually the spade type) for boring through thick walls, timbers, and the like. Two or three shafts can be coupled together for some projects. The size of the shaft is about ⅝-inch, and most sockets accept only ¼-inch bit shanks.

HOLE SAW cuts several diameters.

A bellhanger's bit is an extremely long bit (about 2 feet) that derives its name from its use by electricians in installing doorbell wiring. It functions like an extension shaft but can be smaller in diameter, down to 3/16-inch. Because this long bit tends to whip, run it slowly.

Hole saws cut large neat holes through sheet metal, plastics, plywood, and thin wood. The inexpensive one shown has four circular blades varying from ⅞ to 2½ inches in diameter. Other types cut holes up to 5 inches in diameter. Check a large one before using it; some are made only for use in a drill press.

Large Attachments

Many of the large attachments for a drill are secured to its case. Overall, they perform quite satisfactorily, but they are not as fast or accurate as a tool designed specifically for the purpose. For example, if you intend to build a few storage cabinets in the kitchen, a $10 saw attachment for your ¼-inch drill will serve well for cutting the ¾-inch boards and plywood that you will use. But if you intend to build a garage, you'll be better off buying a $30 to $50 portable electric saw (see page 10). It will cut studs and rafters, as well as thinner materials, more efficiently.

To prevent stalling and overheating, you need a medium or heavyduty drill to drive large attachments. If you intend to buy several attachments, consider choosing a ¼-inch drill with studs or bolts on it that are designed to mate with attachments of the same brand. These are usually better balanced and will be held more rigidly than those with a clamping device designed to fit all drills.

In any case, be sure to take your drill along with you when you shop, making certain that the attachment you buy will fit and will not interfere with the flow of air to the drill's motor compartment.

Following is a list of some of the large attachments:

A drill press attachment is an excellent addition to your drill. It virtually gives you the accuracy and leverage of a regular drill press, and the drill can be easily removed for other uses.

A planer attachment cleanly smooths edges of boards and doors. It is slower than a router's planer.

FIVE-INCH SAW snaps onto a ¼-inch drill. It cuts ¾-inch plywood well, but won't cut through a 2 by 4.

GRINDING WHEEL attached to an electric drill extends versatility, and is less expensive than a regular grinder.

DISC SANDER on a drill smooths wood rapidly; has a ball-joint action on the shaft so it will lie flat.

A right-angle disc sander-buffer, held like a regular disc sander, is less tiring to use than a straight disc sander accessory. Its small discs (with a 5-inch diameter) cut wood at a good speed.

A nibbler attachment chews tidily through sheet metal, including steel. Unlike tin snips, it doesn't curl the edges.

A reversible screwdriver-speed reducer performs well as a power screwdriver and power wrench for small bolts. Geared down 7 or 5 to 1, it also gives your ¼-inch drill extra torque for handling oversized bits.

A circular saw attachment cuts quite well. Since a 5 or 5½-inch blade is the largest a ¼-inch drill can power, the saw does not cut very deeply. For safety, choose one with a blade guard.

A stand transforms your drill into a multi-use bench motor for grinding, sanding, buffing, and the like.

A speed control works very well on any drill and on other small power tools. You simply plug it into the outlet, plug the tool's cord in it, and dial the speed desired. A built-in fuse prevents overloading.

Power Tool Shock Hazards

Many older power tools are neither grounded (earthed) nor double-insulated. Such tools can give a serious—even fatal—shock. For this reason, never use power tools in a damp area unless they are properly grounded or double-insulated. (With cordless, battery-operated tools, there is never any shock hazard.)

Modern power tools may be single-insulated, double-insulated, or all-insulated.

*__Single-insulated tools__ require effective grounding (earthing). Connect them to a grounded outlet through a 3-wire cord and a 3-prong plug. Do not adapt a 3-prong plug to an outlet intended for 2-prong plugs. The outlet box may not be grounded, and adapters are not always reliable. If your electrical system has ground (earth) leads, have the outlet replaced with a grounded-type outlet. Otherwise use double-insulated tools.

*__Double-insulated tools__ are provided with an inbuilt second barrier of protective insulation. Such tools are clearly marked as double-insulated and should not be grounded (earthed). Double-insulated tools are the safest choice for use with older electrical systems where the outlets are not grounded.

*__All-insulated tools__ are marked also as double-insulated and do not require grounding (earthing). They are readily identified because the outer casing is made of insulating material.

For additional power tool safety information, see page 11.

HOW TO DRILL PROPERLY

Three drilling problems often crop up: 1) Centering the moving drill bit on its mark, 2) Drilling a hole straight, and 3) Keeping the wood's backside from breaking away as the drill bit pierces. Following are time-tested techniques.

Keep a pointed tool handy for center punching starting holes. A couple of taps with a hammer on a large nail, nailset, or punch will leave a hole that will prevent the bit from wandering.

Unless you have a drill press or a press accessory for your power drill, drilling straight holes may be difficult. Three methods you can try are shown at right.

To keep a drill from breaking out the wood's backside, do one of two things: 1) Lay or clamp a wood scrap firmly against your work piece's backside and drill through the work piece into the scrap, or 2) Just before the drill pierces, flop the work piece over and finish drilling from the other side.

How do you know when the drill will penetrate? You can either buy a depth gauge made for the purpose or improvise as illustrated below.

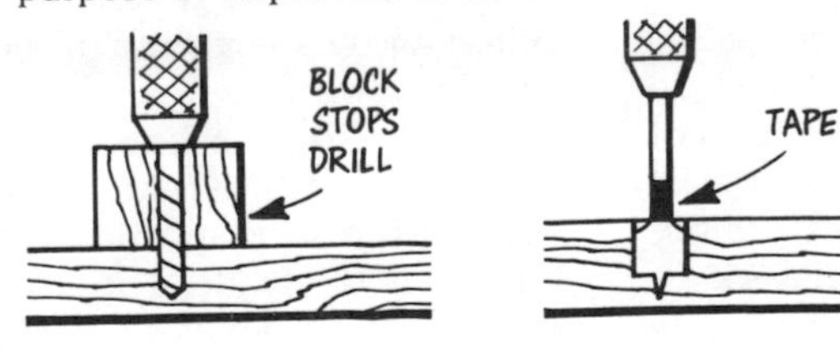

DRILL STRAIGHT with a commercial guide (left) or with one made from a scrap block of wood (right).

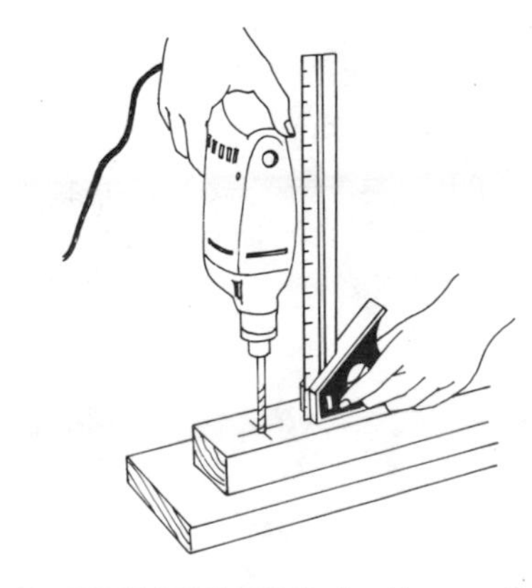

COMBINATION SQUARE helps line up the drill when the holes must be straight.

Tools for Layout and Gauging

The proudest day in your career as amateur carpenter is the first time you finish a project and find that every piece fits. At last you have mastered the difficulties of "true" construction.

Mastering "true construction" calls for skill with cutting tools, of course. But first the lines, angles, or curves must be laid out accurately. For this you need measuring tools, squares and bevels, levels and plumbs, and marking tools.

Although hundreds of different tools are made for precision measuring and marking, basic carpentry requires only a small selection. In fact, for minor projects around the house, a 12-foot measuring tape, combination square, and level should be sufficient. Buy these tools first, adding others as you move into more advanced projects.

Below is a discussion of the primary layout and gauging tools.

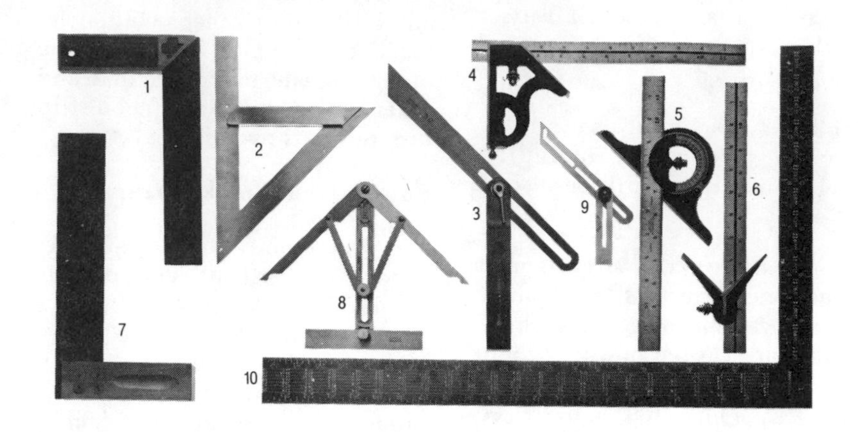

SQUARES AND BEVELS: (1 & 2) two types of combined try and miter square, (3) sliding T bevel, (4) combination square, with special attachments, (5) protractor, and (6) centering head, (7) try square, (8) angle divider, (9) sliding T bevel, (10) framing square.

MEASURING TOOLS

No matter what carpentry job you're undertaking, you'll need an accurate measuring tool. In measuring distances of less than 1 or 2 feet, you can use the blade of a square. Most are graduated in inches and fractions. But for longer stretches, either a flexible steel tape or a folding wooden rule is essential.

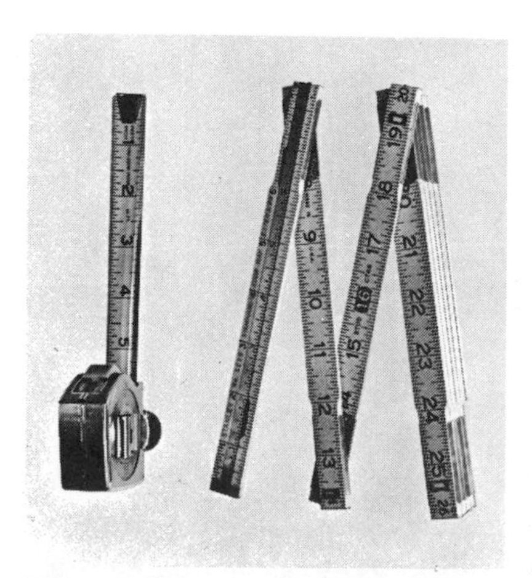

RETRACTABLE STEEL TAPE is one of carpenter's handiest tools; wooden rule is accurate for special jobs.

Flexible steel tapes. Compact and easy to use, flexible steel tapes come in both short and long sizes, ranging from 2 to 100-foot lengths. A 12-foot tape is the one most commonly used. The tape container's base is usually exactly 2 inches long, making inside measuring easy by allowing you to add 2 inches to the tape's reading. With most types, the extended tape recoils automatically into the case, so be sure that your tape has the handy feature of locking into an extended position. The first 6 inches of the tape are usually graduated in 32nds of an inch; later inches are in 16ths.

The end hook of your tape may appear to be loosely riveted to the tape. Actually, it is intentionally made that way, sliding the distance of its own thickness in order to adjust for this thickness. This permits inside and outside measurements to be precise.

Folding wooden rule. Because of its rigidity, a wooden rule will do several jobs that a steel tape can't do. A very accurate tool, it can be extended for some distance without support on the far end. It has a sliding extension to make precise inside measurements and is sharply angled to measure around corners or to aid in drawing angled lines.

Made from several 6 or 8-inch-long wooden sections hinged together, wooden rules generally measure 6 feet. Be careful to open each section in the right direction so that hinges will not break.

SQUARES AND BEVELS

Squares measure 90° and 45° angles; bevels, any angle. Of all the squares and bevels shown here, a few, such as angle dividers, won't often be needed in ordinary construction. However, it is helpful to know that such a tool is available if a particular job requires it. And, as most home carpenters discover, when a new tool comes your way, you'll find ways to make it useful.

Squares

A square must be accurate. If it isn't, throw it away and get another one. To test for accuracy, hold the handle snugly against a straight-edged board and draw a line along the blade. Point the stock in the opposite direction and draw another line along the blade. The two lines should match exactly. Treat a square kindly. Never use it for hammering or prying. Wipe the blade with an oily rag occasionally to keep it from rusting during storage.

If the numbers become hard to read, put a dab of white paint on the square and then immediately wipe it off. Enough will stay in the embossed figures to make them more easily readable.

A try square is available in blade lengths of 3 to 15 inches. You'll often need one for laying out right angles and for testing whether your work is square.

To "try" a board edge for squareness, place the handle firmly along one surface, sliding the blade into contact with the board edge. If light shows between the blade and board, the edge is untrue. Plane or sand the dark ridges until the edge is square and light will no longer show between the blade and board.

The combined try and miter square has a bevel on the handle so that you can also lay out 45° angles. Don't rely on it for a line longer than a few inches.

The combination square is useful as a try square, miter gauge, level, plumb, gauging tool, and 12-inch rule. The tool's freely sliding head can be tightened to the blade or removed. Because the head meets the blade at a 45° angle on one side and a 90° angle on the other, the tool will check for precise 45° miter as well as for square. Most combination squares are equipped with a spirit level for checking true level and plumb.

Two supplementary heads are available: a protractor and a centering head for finding center and diameter of a cylinder.

The framing square is much more useful than it would appear at first glance. Obviously, it has a sturdier blade and can rule longer lines than the try square. In addition, both sides of both blades are inscribed with useful tables and scales.

The framing square is to the carpenter what the slide rule is to the engineer. It serves him as a quick reference when he has to lay out rafters, compute board feet, figure out

TRY SQUARE easily checks board for squareness.

COMBINATION SQUARE (left) does many jobs. (Note special heads at top and bottom of photo.) If light shows under the framing square's edge (right), board is warped.

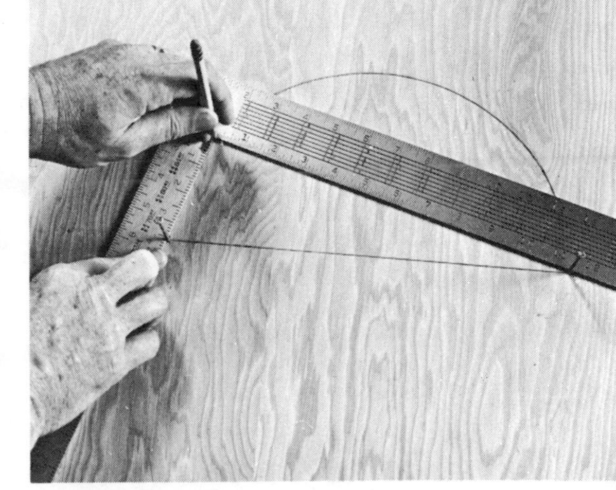

FRAMING SQUARE (left) can help draw square lines across wide surfaces. Drawing a semi-circle is easy (right). Distance between nails equals circle's diameter.

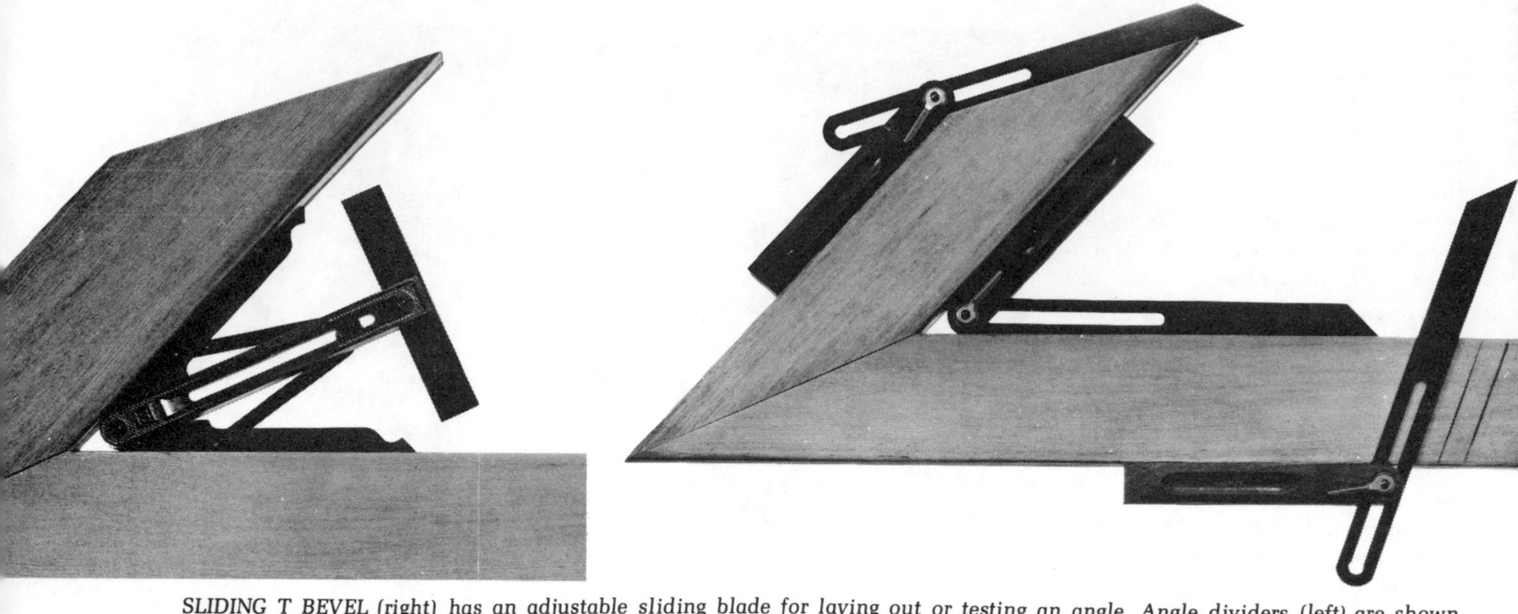

SLIDING T BEVEL (right) has an adjustable sliding blade for laying out or testing an angle. Angle dividers (left) are shown taking off an angle. When set at right angles to the handle, the blade makes the tool a usable try square.

the bracing of post and beam construction, lay out 8-sided figures or other polygons, or measure in 10ths or 100ths of an inch.

The most popular framing square has a 12-inch tongue and an 18-inch body. But if necessary you can get a larger square with a 16 to 18-inch tongue and 24-inch body. These squares come in polished nickel, copper, and blued finishes. Figures are easier to read on a blued square.

Bevels

Bevels are auxiliary tools used most often in combination with a square.

The sliding T bevel resembles a try square but is adjustable to any angle. Use it primarily to transfer different angles from one piece of stock to another. The steel blade has a 45° bevel on one end.

Angle dividers are really double bevels, helpful when you have to divide (or double) an angle in a complicated fitting job. They transfer and divide an angle in one step. With a sliding T bevel, however, this requires three separate steps.

LEVELS AND PLUMBS

In house carpentry a crucial consideration is keeping all horizontal surfaces level and all vertical members plumb. Problems of ill-fitting windows, doors, and wood joinery are often the result of inaccurate leveling and plumbing of surfaces.

The level, as its name implies, enables you to check true level of horizontal surfaces. It also permits checking of vertical surfaces for plumb. Standard levels, 24 inches long, are made of wood, aluminum, or lightweight alloys. At the tool's center, a glass tube holds an air bubble in water. When the lines marked on the tube's surface exactly frame the bubble, the surface is level. A similar tube near each end of the tool indicates plumb.

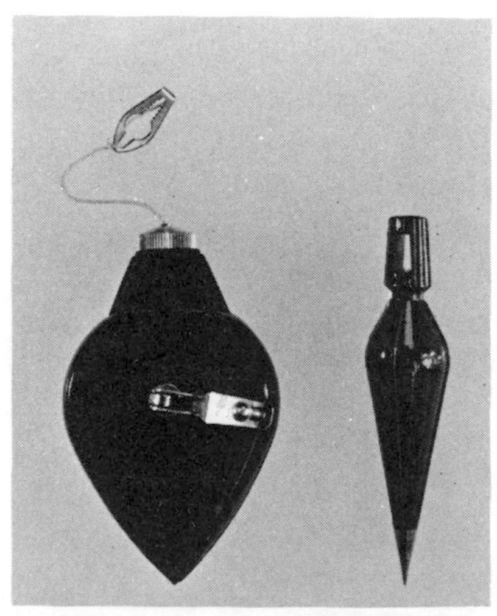

SHARP at one end, plumb bob pinpoints true vertical. Chalk line also acts as plumb bob.

A plumb bob can also be used to check true plumb. This simple device consists of a heavy, pointed weight suspended from a length of string. Holding the end of the string high, with the bob a fraction of an inch above the ground, you can determine true plumb (precise vertical). You can improvise make-shift plumb bobs in several different ways: suspend a pointed weight from a string; buy the special metal plumb bob pictured and attach a string; or use the chalk line shown as a plumb bob.

MARKING TOOLS

A trusty pencil is not the only tool used for marking in carpentry. Several special tools—the chalk line, marking gauge, and wing dividers—are also handy for marking.

The chalk line (shown with the plumb bob) is simply a long spool-wound cord encased in a special container filled with chalk. The chalk-covered string is pulled from its case, stretched taut across a surface, and snapped directly downward so that it leaves a long, straight

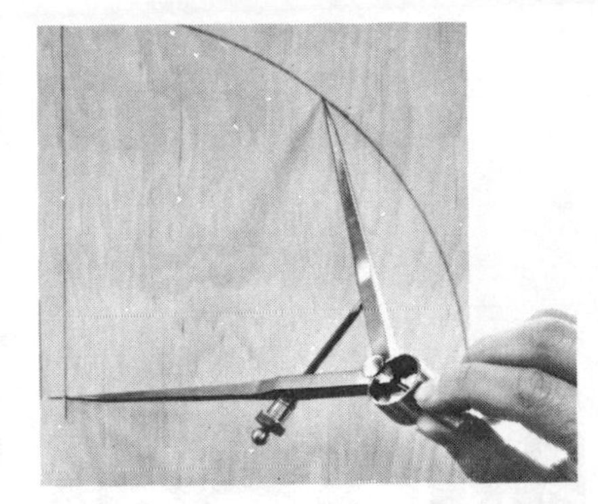

MARKING GAUGE facilitates scribing a line at an exact distance from a board edge. Wing dividers, in photo at right, can be used for transferring measurements or, as shown, to lay out round lines (in this case, rounded corners).

chalk mark. Having an assistant makes the job easier, but most have a hook on one end for single-handed work. Snap the string only once for each desired line.

A marking gauge serves in scribing a line parallel to a board or panel edge. Most gauges consist of a thick, rounded rule that can be tightened down at any particular measurement. A sharp pin at the edge of the rule scratches a mark as the block rides along the board edge. A combination square and pencil can be used for the same purpose.

Wing dividers are used to scribe circles and arcs and to transfer marks. Their appearance is similar to that of a drawing compass. Differences are that both legs of wing dividers have metal points and the distance between their points is adjusted by a knurled screw. With some types, one of the metal legs can be replaced with a pencil.

Clamping Tools

Sometimes having a few good clamps available can be as helpful as an extra pair of hands. They can hold a piece of lumber to a workbench, enabling you to work with both hands, or they can exert extreme pressure, clamping pieces together while they are being bonded with glue.

A carpenter's tip: to protect wood surfaces from being marred by the jaws of a metal clamp, fit a scrap block between the jaws and the wood surface.

Listed below are the major commercial clamps. C-clamps are the most versatile; buy them first. The woodworker's vise and adjustable hand-screws are also handy for most jobs. Special situations may require the use of bar clamps, band clamps, or spring clamps. Buy them when you need them.

C-clamps. Handy for clamping materials either together or to a workbench, C-clamps have jaw widths varying from 3 to 16 inches.

Adjustable hand-screws. Both angle and distance between the wooden jaws of these clamps are adjustable, making irregular flat-sided objects easy to clamp. Keep jaws parallel to maintain even pressure on flat surfaces being glued together.

Woodworker's vise. Its name suggesting its versatility, this handy tool fastens onto your workbench where it will hold many materials securely, freeing both of your hands to work. Choose a vise with large jaws, for they will evenly distribute pressure. Most vises come without hardwood faces; you add the wood with small screws. When bench-mounting a vise, keep the top of its jaws flush with the bench top.

Band clamp. Used for clamping unusually-shaped pieces together, a band clamp consists of a canvas strap loop that is drawn tightly into a metal buckle.

Bar and pipe clamps. Helpful in clamping across an expanse, bar and pipe clamps have adjustable jaws that slide along a bar or pipe. A pipe clamp will reach a considerable distance if you couple several lengths of pipe together. Bar clamps work on the same principle, but jaws won't extend past one bar's length.

Spring clamps. For clamping light work, spring clamps are simple and handy tools. Made like large clothespins, two spring-loaded handles keep jaws clamped tight until they are squeezed together.

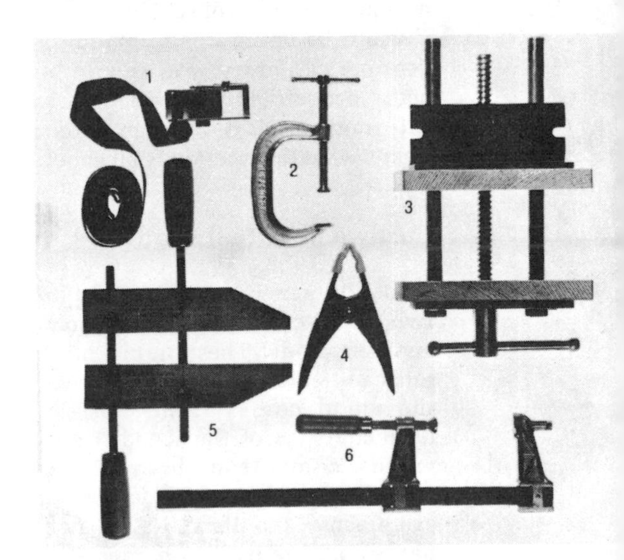

(1) band clamp, (2) C-clamp, (3) vise, (4) spring clamp, (5) hand-screw, (6) bar clamp.

Materials

Though the basic materials of carpentry—nails, lumber, and other building products—may seem pretty matter-of-fact, it's well for a beginning carpenter to remember the old adage, "Big oaks from little acorns grow." With a working knowledge of the basics, you can stretch your skills a long way.

Basic carpentry materials are the subjects of this section. Here you'll find the materials a carpenter will need for general home carpentry work: lumber, plywood, moldings, nails, metal connectors, special fasteners, adhesives, caulking, flashing, insulation, and preservatives. Other materials used only at certain areas of the house, such as wall paneling or ceiling tiles, are discussed in full in their respective sections.

What You Need to Know about Lumber

The endless stacks of different sizes, species, grades, and shapes of wood (and the jargon used to discuss them) can be overwhelming to anyone who wanders into a lumberyard. Of course, the hobbyist who is building a simple bookshelf can close his eyes and buy a couple of boards. But if you'll be doing a fair amount of genuine carpentry, you should be better acquainted with some basic information and lumber lingo. That's what this section is all about.

Hardwood or Softwood?

Woods in general are divided into two broad categories: hardwoods and softwoods. These terms, often misleading, refer to the kind of tree the wood comes from, not the characteristics of the wood. Hardwoods come from broadleafed (deciduous) trees; softwoods from evergreens (conifers). Although hardwoods are usually harder than softwoods, some softwoods—like yew and Douglas fir—are harder than so-called hardwoods like poplar, aspen, and Philippine mahogany. Balsa, the softest of woods, is technically a hardwood.

As a rule, hardwoods are more expensive and harder to work with than softwoods. Because they are often more handsome and durable than softwoods, they are saved for projects requiring these traits, such as furniture or hardwood flooring. Only softwood is discussed here. Information on hardwoods is given only in a section (the flooring chapter, for example) that discusses particular applications of hardwood.

Wood Species

Construction softwoods vary in strength and durability. For load-bearing work, choose Douglas fir or Southern pine. If not available, use larger sizes of other species. Choose redwood or cedar heartwoods where exposed to moisture or weather. When in doubt, ask your dealer.

Lumber Sizes

Given below are some of the most common terms used in lumberyards to describe lumber sizes.

Board foot (Abbreviated as bd. ft., BF, BM): the common volume unit for sawn lumber. Originally the basic unit was probably 1 foot square and 1 inch thick, but the term "board foot" now means any piece of lumber containing 144 cubic inches (for example, a piece 2 by 6 by 12 inches). This measurement, however, is based on the dimensions of rough lumber. Finished lumber is smaller in width and thickness; a finished 2 by 4 is actually about 1½ by 3½ inches. So the term "net board feet" is often used to indicate actual volume of finished lumber. To figure board feet, see inside front cover.

Linear foot (or running foot): a measure that disregards other dimensions and considers only the length; often used for convenience when pieces of the same width and thickness are used in quantity.

Boards: lumber less than 2 inches thick, and from 4 to 12 inches wide (rough).

Dimension lumber: lumber 2 to 4 inches thick, and 2 or more inches wide (rough).

Planks: lumber 2 to 4 inches thick and 6 inches or more wide.

Posts: large-dimension lumber used vertically—generally to support loads.

Strips: a term often applied to lumber less than 1 inch thick and 3 inches wide.

Studs: lumber of a 2 by 3 or 2 by 4-inch dimension, commonly used for framing members in house walls.

Timbers: lumber 5 inches or larger in its smallest dimension; used in heavy construction.

Grading

Information on softwood lumber grades and how to use them is given on the inside front cover. The most important factor in determining a piece of lumber's grade is the type and number of defects it has. Below are descriptions of some possible defects in lumber.

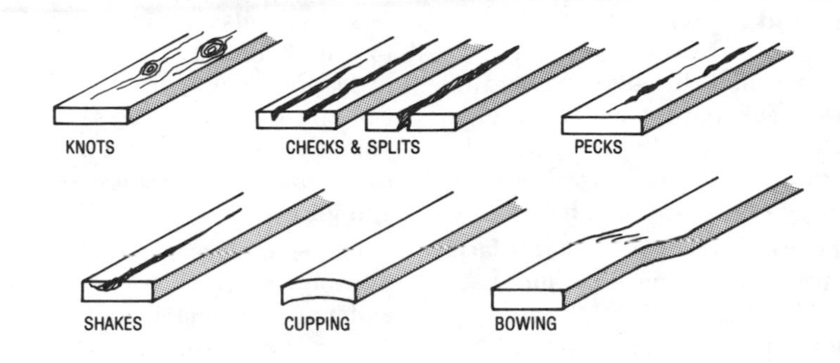

Checks and splits: separations of the wood along the grain. A split extends fairly deep into the wood; a check is a shallow or surface separation and is usually not as long as a split.

Knots: hard (and usually dark) spots of knurled wood occuring where branches once met the tree's trunk. In some cases, knots are loose enough to fall out.

Pecks: areas of disintegrated wood, usually caused by localized decay that was present in the living tree. In lumber that has been seasoned, pecks will not grow larger.

Shakes: separations between growth rings, lengthwise along the wood (the same word is also used to describe a rough type of split lumber used like shingles for outside siding and roofs—page 45).

Wane: bark, or lack of wood at the edge or corner of a piece of wood.

Warp: any variation from a true or plane surface; includes both cupping and bowing.

Your Plywood Primer

With all the plywood produced in the United States in a single year, you could lay a panel 12 feet wide and ⅜-inch thick reaching from the earth to the moon. Yet in spite of plywood's popularity, some home woodworkers don't really understand it well enough to take advantage of its best qualities.

Plywood is a panel made from thin layers of wood called veneers. An odd number of veneers are glued together with the grain of each at right angles to the one adjacent. The grain of the two outside plies is almost punctureproof, and the panel is, pound for pound, one of the strongest of building materials.

TYPES OF PLYWOOD

The type of glue and the grade of the inside plies determine whether plywood is suitable for exterior use. Exterior plywood, with higher-grade panels, has a gluebond that is so resistant to moisture that it will not delaminate even when boiled in water. You can get interior panels with this same glue and with interior panel face grades.

Usual panel dimensions are 4 by 8 feet, but widths vary from 36 to 60 inches, lengths run up to 12 feet. Thicknesses range from ¼-inch to 1⅛ inches. Many stores sell quarter or half sheets of plywood.

All plywood is divided into two main types: softwood and hardwood. Hardwood plywood is most often used for fine paneling and furniture; softwood is the type primarily used for general carpentry.

Hardwood Plywood

You can identify hardwood plywood by the name of the wood used on the face panel. Popular domestic hardwoods include birch, black walnut, cherry, maple, and oak. A number of foreign and exotic woods are also available.

Veneer core, with layers of ply joined in the standard manner, is used for paneling, sheathing, or situations in which the plywood must be bent or curved. Lumber core, recognizable by the solid core or extra-thick middle ply, is used for tabletops or doors where butt hinges are specified. Particle board core, with its core an aggregate of wood particles bonded together with resin, is less expensive.

These are the four most common grades of hardwood face veneer: #1 — flawless, specially ordered, very expensive; "A" — well matched, uniform color; #2 — matching of veneers less careful, pin hole knots; #3 — good if it is painted.

Softwood Plywood

Well over half the plywood sold is made of softwood, usually Douglas fir. Other softwood species include redwood, lauan (Philippine mahogany), cedar, and southern pine.

The appearance of a panel's face and back determine its grade. Letters "N," "A," "B," "C," and "D" indicate the different grades. "N" designates defect-free, all-heartwood or all-sapwood veneers. Use it where you want a perfect "natural finish." It may need to be specially ordered. "A" has neatly made repairs on a surface that can be painted or finished naturally. "B" surfaces may have circular repair plugs and tight knots. "C" plys can have knotholes slightly larger than 1 inch. Limited splits are permitted. A grade known as "C-plugged" has a repaired veneer that allows splits up to ⅛-inch in width and knotholes up to ¼ by ½-inch. "D" grade may have knotholes as large as 3 inches in width and limited splits. For more information on grading, see the inside front cover.

Typical surface textures include standard panels, sanded to a fine finish; surfaces overlaid with a special resin (unsurpassed for painting), resawn surfaces with rough, brushed, or striated textures; and surfaces that have grooves—wide or narrow, deep or shallow, many or few. Many of the panel types combine texture traits.

You can also get factory-finished panels. It usually pays to invest a little more money in order to avoid the labor of finishing the panels yourself. Such panels are available in a wide range of colors, patterns, and glazes.

For decorative usage, you can get plywood panels in a variety of colors and patterns that have been face-veneered with vinyl for extra protection, as well as finishes that resemble marble, wood grain, and tortoise shell. The quality of finishes varies according to the price.

HOW TO WORK WITH PLYWOOD

Using plywood differs from using lumber because of plywood's laminated nature. Here are some characteristics and rules you should keep in mind when working with plywood sheets.

Storing

Plywood should be stored in a cool, dry place. Moisture can crack the smooth finish of softwood panels; exposed edges are particularly susceptible to moisture penetration. The panels should lie flat but off the ground, with adequate support to keep them from sagging. If the wood is covered, allow for air circulation to prevent moisture condensation.

Handle the wood carefully to avoid gouging it or damaging the laminated edges.

Cutting

To avoid waste when cutting several pieces from a panel, sketch the arrangement on a piece of paper before marking the panel. Allow for the saw kerf (the width of the blade's cut) between adjacent pieces. The final cut lines should be marked on the best side of the panel unless you use a portable circular saw or saber saw. If you do, mark the back.

The thin outer layers of the panel can splinter away along the kerf if you cut too fast or if the saw blade is dull. You can make the cleanest, most precise cuts with a saw having a large number of small teeth. Handsaws should have 10 to 15 points per inch, and power saws can be equipped with a special plywood-cutting blade designed for a razor-sharp cut.

If you use a regular power saw blade, adjust it so that just the teeth will protrude through the plywood; this will lessen the blade's impact. Saw teeth should always enter the wood on its good side so that any impact-caused splintering will occur on the back where the teeth emerge. If you're using a portable circular saw, lay the panel with its best face down.

The panel should be supported against sagging. For added protec-

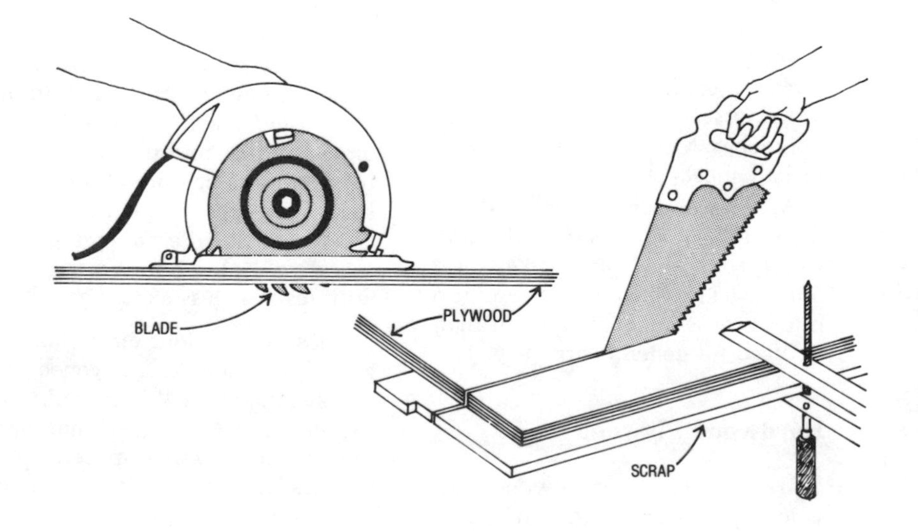

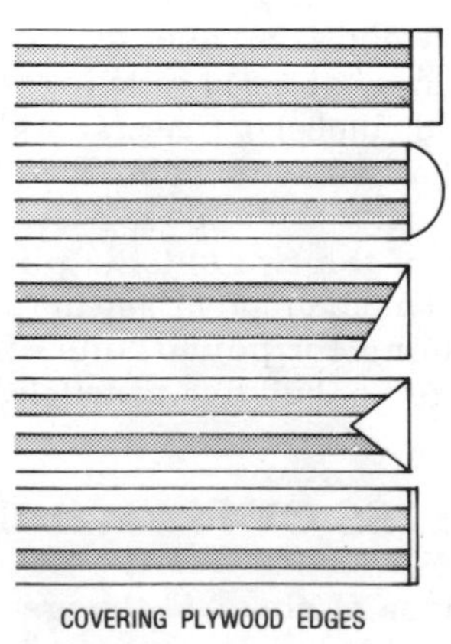

COVERING PLYWOOD EDGES

tion against kerf splintering, put a thin scrap of lumber under the panel and cut it along with the plywood. Feed the wood into a power saw slowly; if you're using a handsaw, hold it at a low angle. If the kerf binds your saw, hold it open with a screwdriver blade.

Planing and Sanding

Careful cutting should give you a fairly smooth edge. But if you must plane the edge of the board, use long strokes with shallow cuts. Work from both ends toward the middle to avoid overplaning or chipping away the ends.

Only the edges should need finish sanding. The panel's surface is smooth and you risk removing the soft grain by sanding it. To fill gouges or nail holes, apply wood putty slightly higher than the panel's surface, let it dry, then sand it smooth. If you sand the surface after applying a finish, work with the grain using gentle strokes with number 120-grit or finer paper. To avoid biting through the finish, use a sanding block.

Fastening

Fasten plywood joints either with nails or using flat-head screws in predrilled holes. Add glue for strength; liquid resin (white) or urea resin glue for interior use, resorcinol types for outdoors. Put a preliminary coat on the edges for absorption, then apply the final coat to both sides of the joint and fasten them together before the bond forms permanently.

Most common joints can be made with plywood, but simple butt joints using panels less than ⅝-inch thick should be reinforced with wood framing on the inside. You can nail near the edge of plywood without splitting, but nailing into the edges of thin pieces is difficult.

All exterior plywood edges, whether or not exposed, should be lightly sealed with a high-quality exterior oil-base house paint primer or similar sealer.

Moldings...for Fix-up

Look around at the walls in one of your rooms. If your house is typical, moldings probably run around the perimeters of doors and windows, along the base of the wall, and possibly along the wall where it meets the ceiling.

Moldings serve two main purposes: they're decorative, and they hide inaccuracies in joints between materials.

The sketch on the next page shows some of the typical molding patterns and areas in a room where they are applied. Shown in the photographs are methods of nailing, cutting, and joining moldings that will result in a professional-looking job.

Since carpenters generally prefer to use moldings only where necessary to cover misfitting gaps between materials, the absence of molding may be the mark of good craftsmanship. Often times, however, there is no getting around the need for molding. And when applied properly, it can look handsome.

Here are some tips on working with moldings.

How to Cut Moldings

The most commonly used tool for cutting moldings is the miter box (shown in the photograph). Using a miter box, you can precisely cut the 45° angles used for joining those

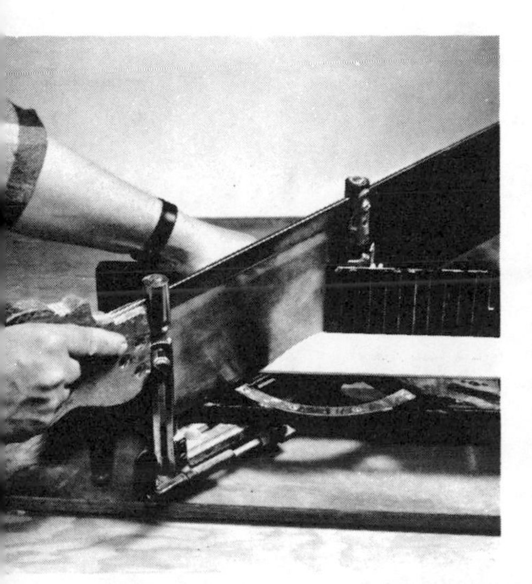

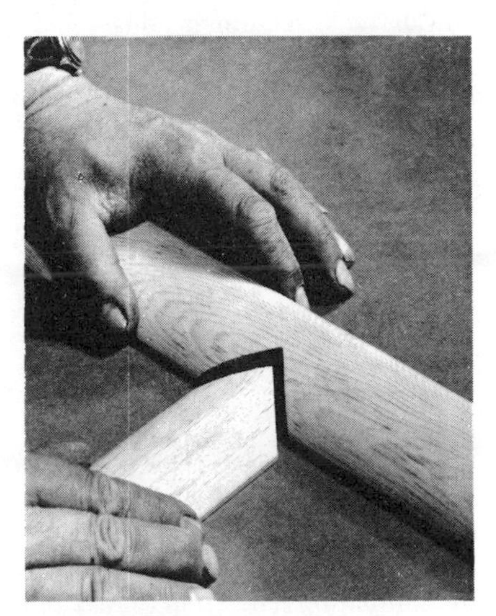

MITER BOX is most commonly used tool for cutting moldings. Here it is shown cutting an exact 45° miter for a corner joint (using door and window molding). For cuts midway along a piece, you'll have to use a fine-toothed crosscut saw.

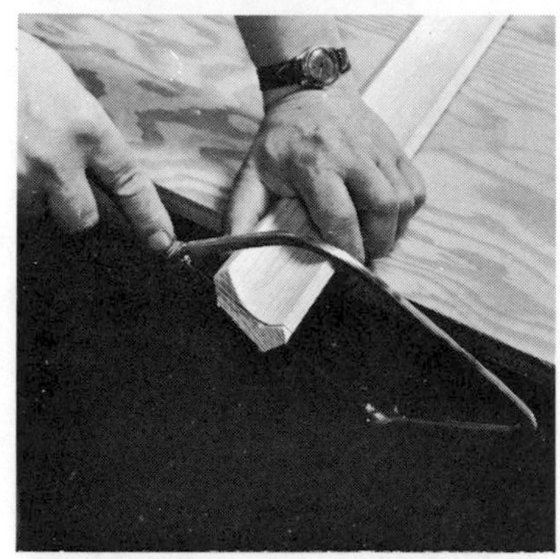

COPING SAW is used to cut proper curve for joining two lengths of three-dimensional-type molding.

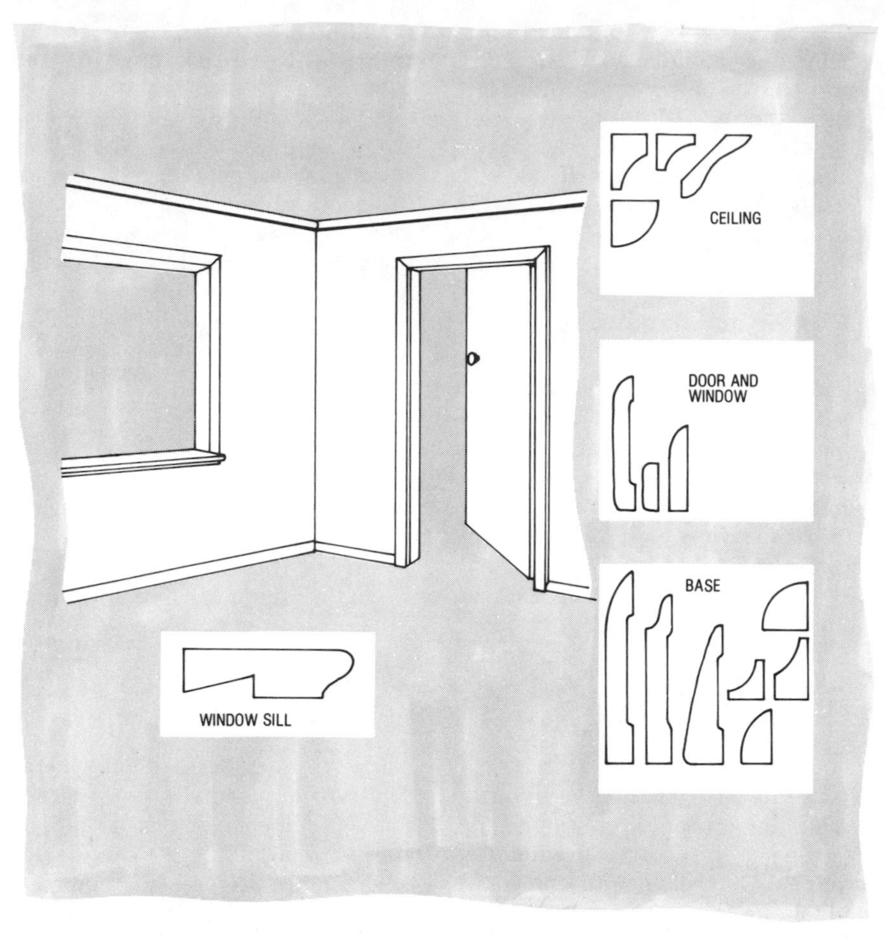

types of moldings applied to flat surfaces (such as around windows and doors). Pointed miter cuts, used to join the end of one molding midway along another length, must be made using a fine-toothed crosscut handsaw or backsaw.

Curved, three-dimensional moldings placed along cornices and baseboards must be mitered with the miter box and then cut to the proper curvature with a coping saw. This process is illustrated in the photographs above.

How to Blind-Nail Moldings

Two methods are used for nailing moldings into place: nailing with finishing nails, then setting the heads with a nailset; and blind nailing. To blind nail, use a small knife or gouge to raise a sliver of wood that's large enough to hide the head of a finishing nail. (Don't break the sliver completely from the molding or you may not be able to replace it in its natural position.) After pulling the sliver to the side, nail into the cavity with a finishing nail, then glue the sliver back into place. You can tape the sliver down with a piece of masking tape until the glue is dry. Rubbing the spot lightly with fine sandpaper will remove all signs of fastening.

Rather than going through this entire procedure for one nail at a time, do it one step at a time, making all the gouges first, then setting all of the nails.

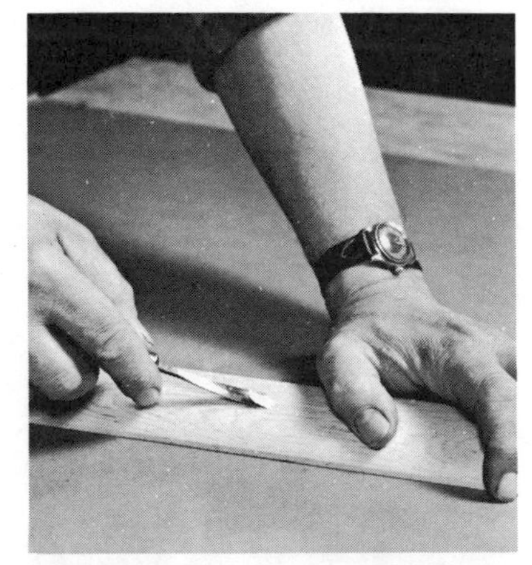

BLIND NAILING molding will hide all signs of fastening. Gouge up a small sliver of wood and nail within the gouge, using a finishing nail. Set the nailhead slightly, gluing the sliver back into place. After light sanding, all signs of a nail are gone.

Ordinary and Extraordinary Nails

The word *nails* once meant only simple iron pins with heads. But today it covers a wide selection of metal fasteners, many with greatly improved holding power, ease of use, and appearance.

You'll find this new, wide choice of nails a great help whenever you're repairing or remodeling at home, for in this type of carpentry you are usually attaching something to wood that has become dry and hard or replacing ordinary nails that have failed.

Shown here is a collection of special nails and of good ordinary nails suitable for general construction use. You can obtain most of these at a hardware store, though for some you may need to go to a builder's hardware supply or lumberyard. These are only a sampling of the nails available today; at least a thousand kinds, counting all sizes and types, are being made.

For basic information on nails and a chart of their sizes, see inside back cover.

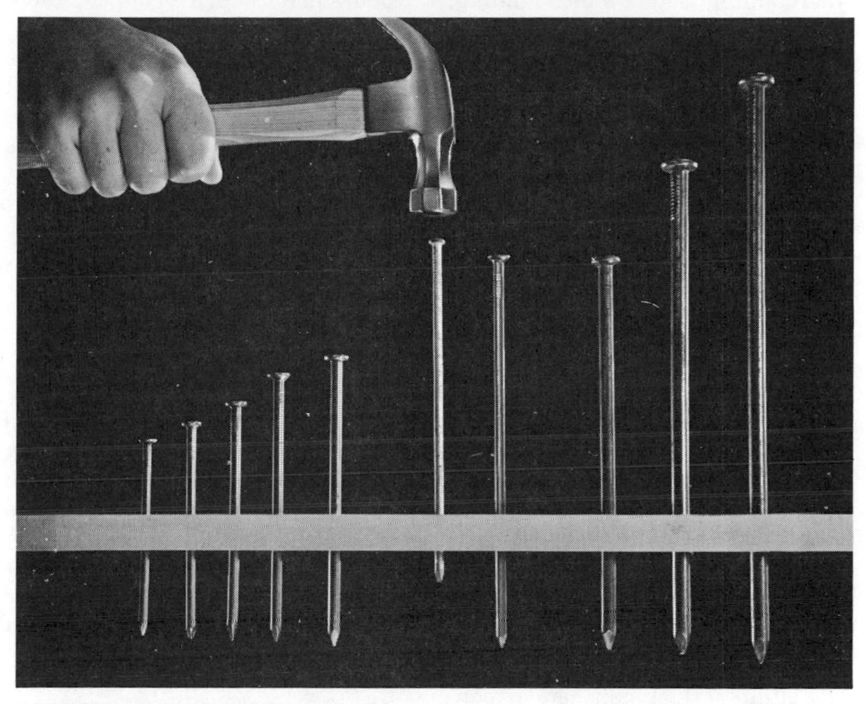

LARGEST NAILS generally available are shown here. Left of the hammer are 20, 30, 40, 50, and 60-penny ("d") common nails. Being hammered is a 7-inch gutter spike; next, a square-shanked decking spike. At right are common wire spikes, 8, 10, and 12 inches long.

Ordinary Nails

You've probably used either common nails or box nails many times. They resemble each other, except that the box nail is thinner. Though it is less likely to split the wood, it bends more easily if you are not adept at hammering.

Choose either the casing nail or the finish nail for finish work. A countersunk finish nail leaves a slightly smaller hole to fill; a casing nail is slightly heavier and stronger.

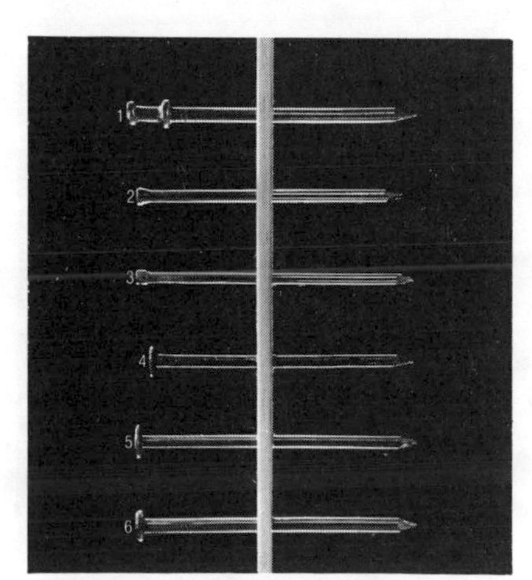

ORDINARY NAILS come in "bright" and galvanized: (1) scaffold, (2) casing, (3) finish, (4) coated sinkers, (5) box, (6) common.

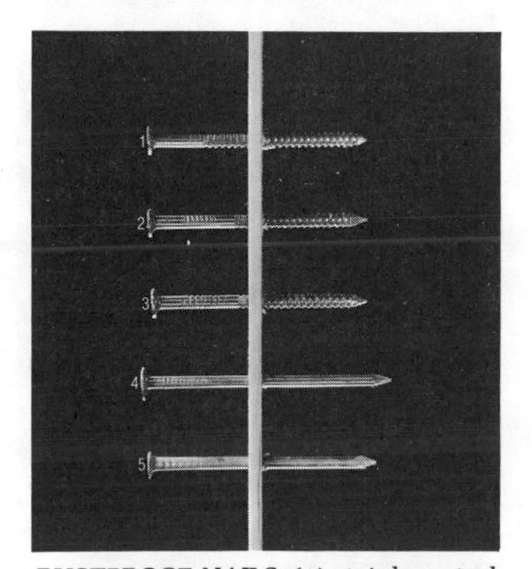

RUSTPROOF NAILS: (1) stainless steel, (2) Monel metal, (3) silicon bronze, (4) copper, (5) aluminum. 1 and 2 have threaded shanks.

Cement or rosin-coated sinkers are widely used to attach and repair gypsumboard walls and ceilings. They hold better than bright nails, and their heads are slightly beveled on the underside to ease sinking them flush.

The scaffold nail is handy whenever you build a structure that is to be taken apart later. Drive the nail down to its inner head, leaving the outer head exposed for ease in pulling it out.

Common, box, casing, and finish nails are available in most stores in sizes from 2-penny to 16-penny; you can usually find the common nail in sizes up to 60-penny (see photograph). Long ago the term "penny" (abbreviated as "d") meant the cost of a hundred hand-forged nails— for instance, a hundred of the smallest nails sold for two pennies. Now, however, the term indicates a nail's length.

Nails larger than 60d are called spikes and are designated by their diameter and length in inches. Unless you are building a railroad-tie retaining wall or a boat landing, you'll probably have little use for the ⅜ by 12-inch spike shown.

Rustproof Nails

It's wise to use either nonferrous or stainless steel nails on any wood siding or screens on the exterior of your house. The best hot-dipped galvanized nail will rust in time, particularly at the exposed nail head, because the coating there is battered by your hammer. Aluminum nails are well suited to exterior work and are an inexpensive answer wherever many rustproof nails are needed. Although they cost more per pound than steel nails, you receive more nails per pound. And, although they bend easily, recent improvements in alloys have made them stiffer.

Stainless steel nails are expensive, but they are often used on exterior siding, especially prefinished siding. Made of tough alloys, the new types are thin-shanked, again providing more per pound and thus lowering the unit cost. They neither rust nor corrode from chemical action. If your house is a few years old and some of its siding has loosened, stainless steel nails are much more satisfactory than aluminum ones for driving into the toughened wood.

If you are making a wood planter box for a water plant in your garden pool and the pool has copper plumbing, use copper or bronze nails. They will not rust, nor will they be affected by electrolysis. If the pool has galvanized plumbing, use stainless steel nails to reduce the electrolysis.

Similarly, use the small aluminum edging nails to attach aluminum edging to a counter top or to furniture.

Masonry Nails

Masonry nails are available in a wide choice of sizes for securing thin and thick materials. You can drive them into concrete, concrete block, soft stone, brick, and other masonry. When driven into seasoned concrete or hard stone, these high-carbon nails tend to chip out the material and then, though they maintain lateral holding power, they have little holding power on any outward pull. In many situations this doesn't matter, but when you are hanging something on a hard masonry wall, drive these nails into masonry joints, drill lead holes for them, or use masonry anchor bolts (next page).

Roofing Nails

You'll notice that roofing nails vary considerably in appearance. They are made to secure various roofing materials, such as asphalt, aluminum, fiberglass, and asbestos-cement. Always ask for a roofing nail designed for the roofing material you are using. Some nails have neoprene or lead washers under their heads to seal the holes they make through the roofing material. The barbed plywood nail included in those shown is used to hold plywood roof sheathing down.

Shingle Nails

Shingle nails (not shown) are made today in special lengths and head sizes for various roofing uses, such as with new shingle roofs, new shingles laid over old, asbestos shingles, and shakes of various thicknesses. To replace a few shingles yourself, you can use the smaller sizes of box nails.

Small Nails

Wire nails, brads, and tacks have changed little in recent years, but there are some new and handy small nails available, as well as a "jumbo tack" (a square-headed nail made for asphalt roofing). It is excellent for securing canvas, plastic sheets, hardboard, or plywood wherever maximum holding power is needed.

The pin nail is a long, thin nail made for securing insulation board and acoustical tile. It's also valuable

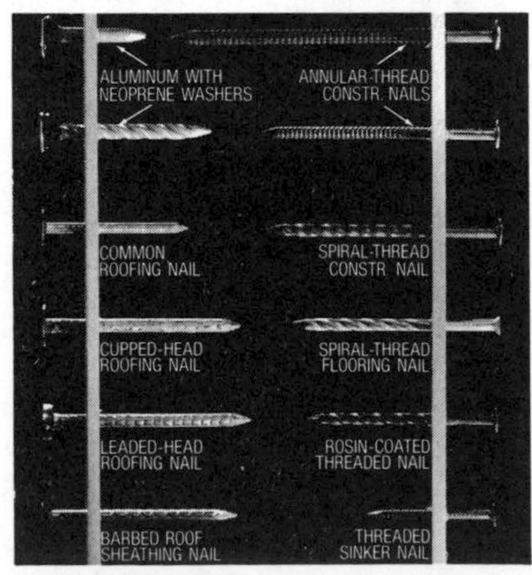

ROOFING NAILS are at left, made for a variety of materials. General construction nails at right have exceptionally good holding power.

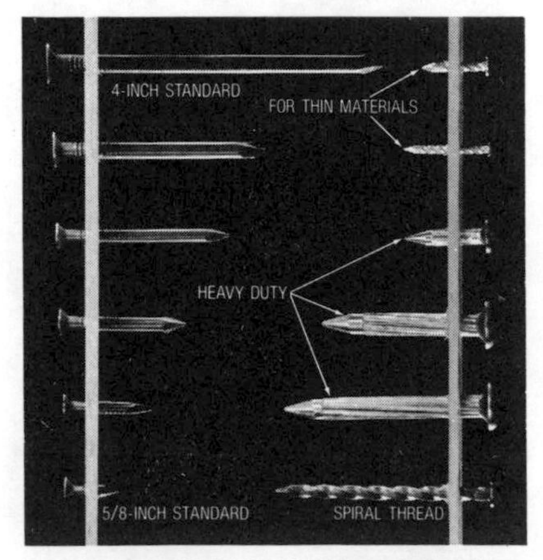

MASONRY NAILS. You can drive these carbon-steel nails into most masonry with a hammer; you must drill lead holes in hard masonry.

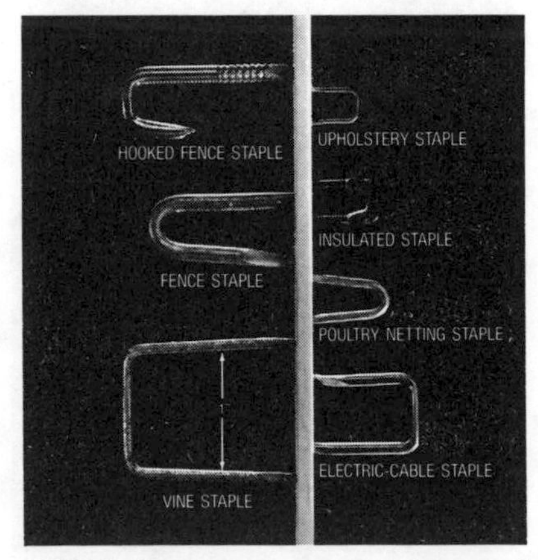

STAPLES come in many sizes and shapes. You can use the types shown here (driven with a hammer) for more purposes than their names indicate.

for assembling a box or cabinet of thin wood.

The acoustical tile nail has its head part way down its shank. When you drive the end of the nail flush with the tile, you cannot see its head.

Staples

The upholstery staple shown in the photograph comes in several small sizes and is widely used for attaching small screening, as well as for upholstery work. The hooked fence staple makes building wire fences easier (you drive the staples partially in the posts and hook the wire over them before stretching it). The vine staple and electric-cable staple are handy wherever you need a large staple. These are often tacked to a shop wall to hold small tools.

Threaded Nails

As you've probably noticed in these photographs, a great many of the new nails have threaded shanks—spiral, annular, or screw thread. Any of the three kinds gives more holding power than galvanized, coated, or barbed nails. They also have two to three times the holding power of smooth nails, particularly after some time has passed.

Until recently, threaded nails were expensive to produce, but because of manufacturing improvements, their cost has been reduced and they are now available even as "common" steel construction nails.

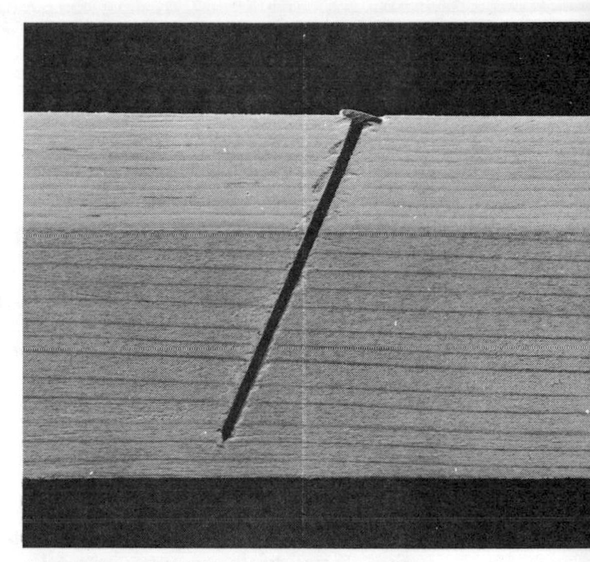

FOR BEST HOLDING POWER, a nail's length should be three times the thickness of board being secured; nail should be driven at a slant.

Fasteners for Non-Wood Surfaces

Wood is resilient, clinching tightly around nails, screws, or whatever you drive into it. On the other hand, plaster, gypsum wallboard, concrete, and other such materials lack the necessary resiliency to hold common fasteners, so you need special fastening devices.

Two separate series of fasteners are designed for securing to inelastic materials. For gypsum or plaster, fasteners depend upon a spreading frame that distributes weight more widely than a nail or screw. Fasteners for masonry use resilient sleeves that fit into drilled holes. In some cases, concrete nails can serve as fasteners for masonry surfaces (see facing page).

Gypsum or Plaster Fasteners

Whereas concrete fasteners may secure objects to floors as well as to walls, gypsum or plaster fasteners are used primarily for hanging objects on walls. The type of fastener to use in one situation depends primarily upon the weight of the object being fastened.

For objects weighing two pounds or less, a straight pin driven downward into the wall at a 45° angle is reliable, durable, and disturbs plaster the least.

The conventional picture hook, a nail seated in a metal hook, should hold up to 20 pounds. To minimize breaking away plaster or gypsum, drill a guide hole for the nail. Use a fine bit (drill with the hook as a guide to assure getting the hole at the correct angle).

Shelf brackets and other weighty objects should be secured by wood screws driven into the wall studs. (See page 54 for methods of locating studs.) In places where you can't drive into a stud, choose screws with expansion anchors or split-wing toggle bolts. With either type, drill a hole, insert the screw (or bolt), then tighten it to spread the anchor (or wings).

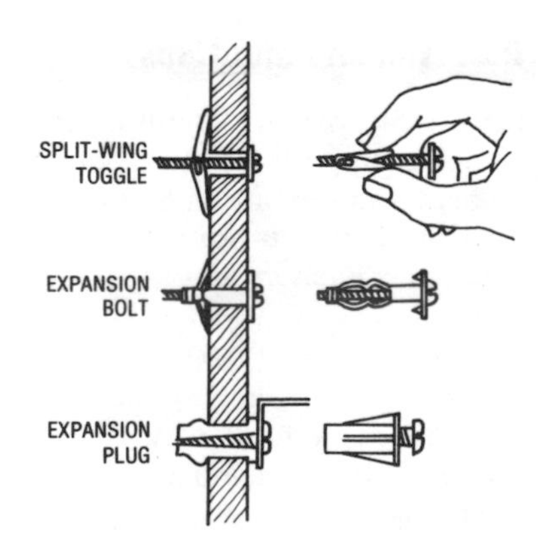

The most frequent mistake made during installation is to turn the screw too far; this pulls the anchor into the wall material. The deeper it cuts, the weaker the device's grip.

If design permits, use two or three anchoring devices to secure a strip of wood to the wall, then fasten the shelf brackets (or whatever you are mounting) to the strip of wood.

When purchasing these anchors, be sure to get the proper size. Expansion anchors, for example, have a solid shank designed to equal the thickness of the wall.

The design limit for a single anchor is 100 pounds of dead weight. Anchors are not intended to secure shower grab bars or anything meant to take sudden pulls.

If one of the devices is abandoned, remove the screw (if possible) and sink the anchor shield into the wall deep enough for it to be covered with caulking compound or a spackle patch.

Masonry Fasteners

A wide assortment of expansion shields are available for securing screws in masonry. Length and diameter govern the carrying weight of these fasteners.

Expansion shields, like fasteners used in gypsum or plaster, can eventually work loose. If this occurs, they should be abandoned and a new device inserted in sound surface.

The key to a good installation is proper drilling of the hole to receive one. Three tools are used: star drills for small, accurate holes; electric drills for faster work; and electric hammers for big fasteners.

Using a star drill takes patience but is a simple matter. You mark the exact point of the hole and hold the drill firmly while making the first five or six hammer taps. Once the hole begins to form, loosen your grip and let the drill tip dance a little as you strike it.

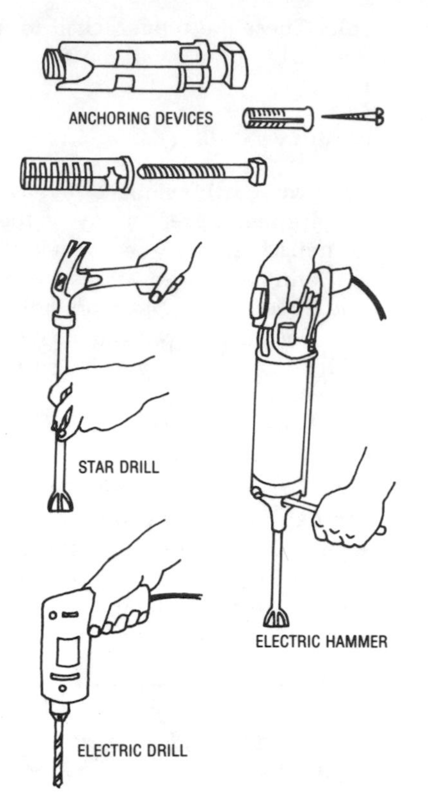

Star drill diameters run from 3/16-inch to 2 inches. For a hole larger than ¾-inch, first drill a pilot hole, then the properly-sized one. In all cases, hundreds of light hammer taps rather than dozens of crushing blows result in a truer hole and fewer cracks.

Electric hand drills with carbide tips are faster and easier to use than star drills but are generally limited in the size bit they'll drive. Although the drills can be rented, carbide-tipped bits must be bought.

For a hole ¾-inch or larger, an electric hammer is the required power tool. It turns a bit and hammers at the same time. This, too, can be rented.

Once the hole is drilled, push the shield into it, then insert the screw through the bracket (or whatever you are fastening).

Metal Framing Connectors

If toenailing and driving large spikes isn't your favorite pastime, you may be pleasantly surprised by the wide selection of metal framing connectors available.

When framing a building, a carpenter often uses quite a few of these small galvanized-steel connectors because they save man hours and assure strong joints.

Simplifying any framing you do, they eliminate the complications and weak joints that often occur where you add on to an existing structure. Most require only nailing in place.

Builder's hardware stores and most lumberyards stock these connectors in a plentiful and somewhat confusing array of types and sizes.

Nails are supplied with many of the connectors. These are heavy barbed or ringed nails, usually 1½ inches long, that fit tightly in the holes in the metal and grip the wood firmly. Using the proper nails is important. If nails are not included, obtain those specified by the manufacturer. Common nails large enough to hold properly will be too long for most connections and will possibly split the wood members.

These main types of connectors are discussed below: post anchors and caps, joist hangers, and universal framing anchors. Special plywood clips are also included.

Post Anchors and Caps

Structural posts for anything from fences to supports for patio roofs can be held securely at the bottom and easily locked into beams overhead using post anchors and caps.

Anchors are easy to "set" and will keep posts from lifting in a wind as well as from moving sideways. They are made to fit 4 by 4's, 4 by 6's, and 6 by 6's (surfaced lumber).

The first anchor shown is for securing posts to an existing concrete slab. Drill a hole in the concrete for an expansion plug and lag bolt (above) to hold the anchor. The anchor has an eccentric arrangement that allows you to move it a bit sideways with the bolt loosened for exact alignment.

The other post anchor is for a new slab or fresh concrete footing. Simply push it down into the wet concrete to its middle partition where it serves as a moisture barrier for the wood post.

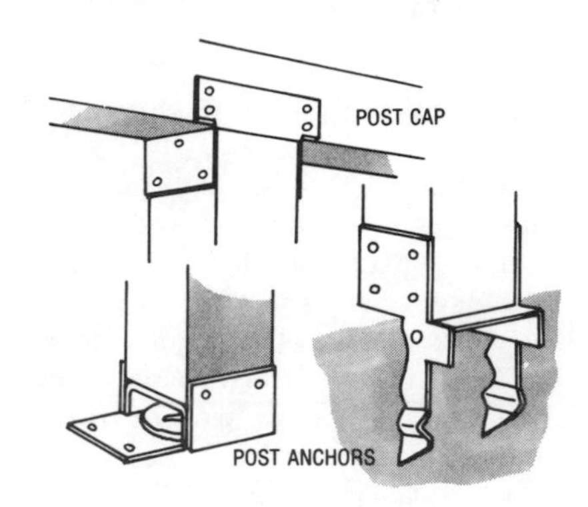

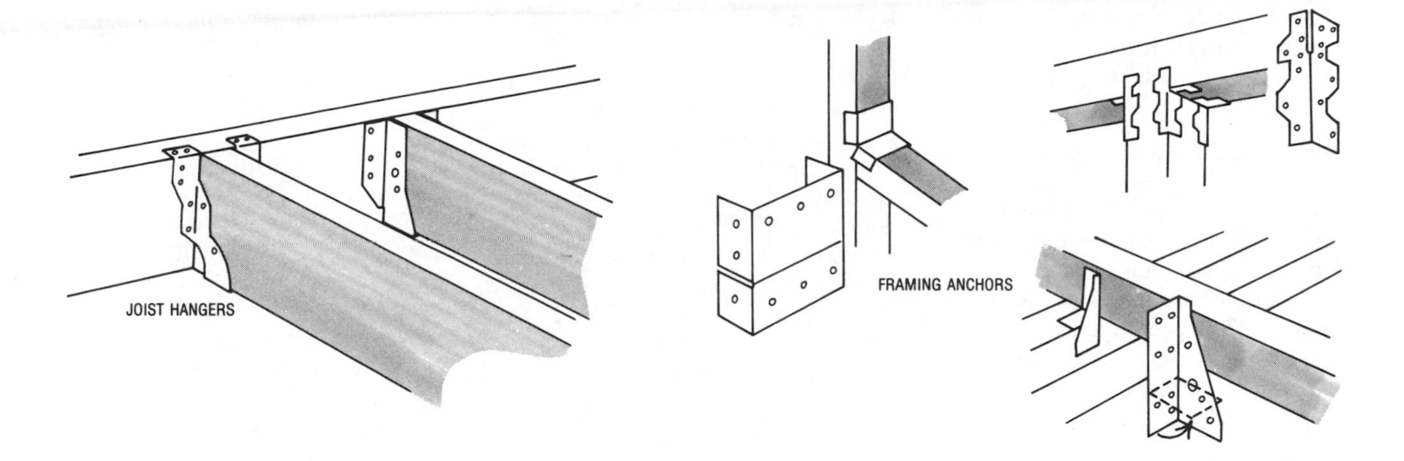

Using metal post caps, as shown, locks the beam to the posts much more securely than toenailing and much more easily than bolting would do.

Joist Hangers

Helpful in securing floor, deck, or roof joists to beams, joist hangers like those shown can save you a lot of time and trouble. Because they fit around joists, rafters, and beams, they come in sizes that fit almost every stock lumber from 2 by 4's to 6 by 6's.

The first type is used to connect a joist directly to a beam. Simply nail the hanger to the beam and nail the joist into its "saddle."

The second hanger shown is for attaching joists to flat surfaces, such as when you attach a patio roof directly to the outside of a wooden wall. Nail the hanger into wall framing or to a ledger board.

Framing Anchors

Framing anchors are the most universal of connectors. Many styles are available; three are shown.

In some cases, you'll need a right and left-hand type to make all connections. Some have one bent end, but other anchors are left straight so that you can bend an end in a vise whichever way desired. The latter are best when you need only a few of each kind for repairs or simple framing.

Plywood Clips

Although they are not in the same class as other heavy-duty anchors and hangers, plywood clips can also be very handy metal fasteners.

They are designed to connect plywood roof sheathing panels along their edges when the joint is not supported by rafters. You can

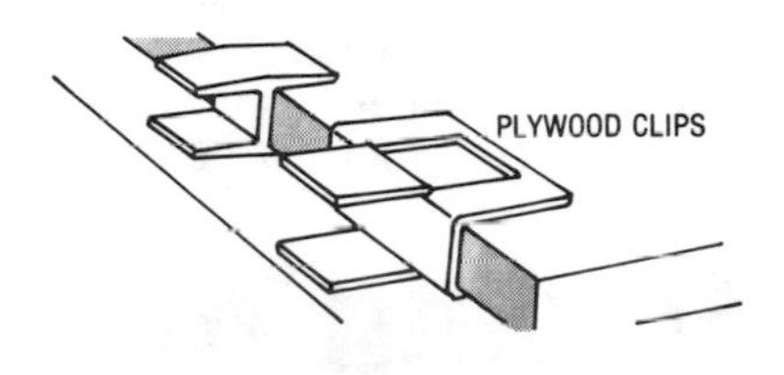

also use these clips to connect plywood or other panels together without the need of a framing member at every joint.

Clips, made of aluminum or steel, come in sizes fitting 5/16 to 3/4-inch plywood thicknesses.

Which Adhesive Is Best?

Adhesives are designed to make two surfaces adhere tightly together. Although their generic name is "glue," they are also sold as "cements" and "sealers." Glue types have proliferated since the advent of plastic and epoxy. Great as the epoxies are, they should be approached with caution on one score: be sure that you want the joint to last forever before applying epoxy. Joints that might require later dismantling or adjustment should be secured with less robust glue.

Caulks and sealants (see page 34) differ from adhesives in that adhesives are not designed to fill gaps.

To make the best glue bond, follow manufacturer's directions exactly. For instance, when gluing porous materials, the directions will often advise you to spread a coating on each surface, let it dry, and then spread second coats before you join the pieces. If you omit the second coat, the material may absorb most of the glue, leaving the joint "starved." Most glues will not bond at temperatures below 70°. The life of some glues is shortened by high temperatures and drying time is speeded up.

In order for any glue to bond properly, the surfaces to be joined must be clean, dry (except with epoxies), and as close-fitting as possible.

Wooden surfaces that do not match can be bridged to some degree by mixing sawdust with the glue.

Animal or Fish Glues

The earliest of woodworking glues, animal or fish glues are most satisfactory for use indoors where temperature and humidity do not vary widely. They can be loosened with heat and water for subsequent adjustment or regluing.

White Glue

Known also as polyvinyl glue, white glue is generally useful, especially for porous surfaces. It sets quickly, does not stain, and gives a slightly resilient joint. Application and limitations are about the same as for animal or fish glue. Because it can be used on paper, plastics, and other miscellaneous materials that animal glue does not bond, white glue is handy to keep around the house.

Resorcin Resin

Although expensive, resorcin resin makes a very strong waterproof joint. Resorcinal glue is often used in building boats and is one of the most durable glues for general outdoor use. It will hold on oily and resinous woods. To prepare it, mix powdered catalyst with liquid resin. You have about 10 minutes to form the joint after applying the glue. The liquid resin is flammable, leaves a conspicuous dark line when it dries, and tends to show through paint.

Epoxy Resin

Commonly used on metal, concrete, and other non-porous materials, epoxy is very strong. Available in a slightly flexible form, it is good for bonding an infinite number of materials but will dissolve some plastics. Epoxy will set at lower temperatures than other glues and in wet conditions. You do not need to clamp. Curing time ranges from 12 to 24 hours.

Contact Cement

Handy for bonding surfaces that can't be clamped, contact cement forms a permanent bond on contact. Since strong-bonding contact cement is usually used on large surfaces, such as for applying counter tops or wall paneling, it is generally applied with brushes or rollers.

Both lacquer-base and non-flammable water-base types are on the market. Use lacquer thinner to clean brushes and rollers for lacquer-base cement; soap and water will clean up the water-base type.

Because the bond formed is permanent, pre-fit all surfaces before applying cements. Apply two coats to each surface and allow each to dry thoroughly. Drying time is 30 minutes to 2 hours. Once cement is dry on both surfaces, carefully align parts and push them together.

Sealing Your House

The primary purpose of a man's house is to protect him from the elements. He lives in a house so that he won't get rained on, snowed on, windblown, or sun baked. In addition to protecting its inhabitants, a house must be able to withstand weather damage.

This section deals with the different steps a carpenter should take during projects to insure a house's ability to keep weather out and comfort in.

Flashing is discussed first. A carpenter working on a roof or exterior wall that is exposed to excessive water must bear in mind the need for flashing.

Caulks and sealants are mentioned second. Almost every job dealing with roofs or walls calls for the use of some type of sealant.

Third, insulation and vapor barriers, to be installed within walls, ceilings, and sometimes under cold floors, are explained.

Last (and very important for any wooden structure contacting concrete, earth, or constant moisture) is a discussion of preservatives and wood sealers.

FLASHING

An important part of roofs and exterior walls of houses, flashing is necessary wherever a flow of water would otherwise penetrate. Usually shaped from flat sheets of non-corrosive metal, flashing protects roofs around chimneys, plumbing stacks and vents, valleys, and areas where any surfaces, like dormer walls, meet or extend through the roof. On walls, flashing directs water off of exterior door and window frames.

Like most work on houses, flashing must meet local code specifications before beginning work. If you find that certain flashing jobs are beyond your skills, you should obtain estimates from a roofer.

For do-it-yourself jobs, you'll encounter three main types of flashings: 1) continuous flashing for valleys where two roof surfaces meet, 2) step flashing, to be laid at the ends of shingle courses (in the same overlapping fashion as shingles) where roofs butt against walls and other surfaces, and 3) special flashings for above doors, windows, and around objects on roofs, such as vents and chimneys.

In the sketch, you can see the areas on a typical house where these types are used.

CAULKS AND SEALANTS FOR WALLS

Even small cracks in your house's exterior walls can bring a host of troubles: uncomfortable drafts, higher heating costs, damaging moisture penetration, and invasions of shelter-seeking insects. Large,

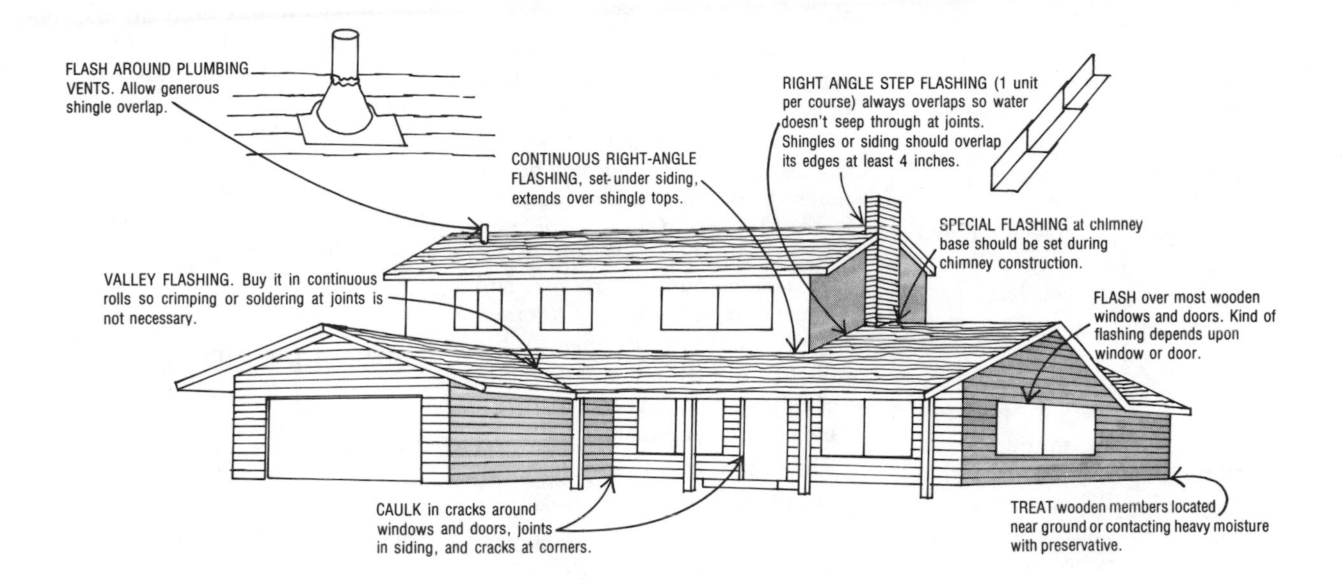

rapidly widening cracks may indicate foundation problems that call for professional help, but you can easily seal most small cracks yourself with commercial sealant or caulking compound.

What Types Are Available?

A little knowledge can help you choose the right one of the many sealants on the market. Ordinary, low-cost sealants will do for most caulking jobs. But for a particular problem, you may need a special compound.

Most caulks are purchased in 11-ounce cartridges that fit standard caulking guns. Some types, known as "rope caulk," are sold coiled. You unwind the rope-like caulk from its spool and press it into the crack with your finger.

Before using a compound, read the label for any special instructions. Some sealants won't stick to certain surfaces without priming, and most have a minimum-use temperature. The label should also note the compound's "tack-free" time, following which it can be painted. While some sealants are manufactured in colors, painting it yourself after application when it has become "tack-free" allows more certain color matching and extends the life of most compounds.

Oil-based compounds, the most common and inexpensive, are adequate for most crack-filling chores. But they're relatively temporary, lasting from one to eight years, and will eventually shrink and flake away. They can be used for most interior cracks, such as around door and window frames and in walls. Don't use them in joints that will move or expand.

Acrylic latex sealants adhere better and shrink less than oil compounds and are easier to use; you can wipe away any excess with water. They resist cracking and embrittlement and last up to 10 years. Although they can be applied to damp surfaces, they shouldn't be exposed to constant moisture. Unlike oil caulks, they won't stain or "bleed into" oil-impregnated woods, and they stick better to tile, glass, old paint, and plaster. Don't use them on metal, in moving joints, or for cracks wider than ¼ inch.

You apply a convex bead of the caulk and then smooth it with finger pressure. Light sanding will give it some texture if it dries too smooth to match a coarse surface, such as that of plaster.

Butyl rubber caulks are flexible, more adhesive, and last longer—up to 20 years. They can be used between dissimilar materials or where some expansion or movement is expected. Butyl is excellent for window glazing, for metal, plastic, and masonry joints, and for cracks between the fireplace and the wall.

Butyl weathers well and resists moisture, but it shouldn't be exposed to constant moisture. Because it shrinks, it is best used in filling narrow cracks. Paint it to keep dust away.

Butyl caulk is available either in tube cartridges that fit in caulking guns or in cans, from which it's applied with a putty knife. Using a putty knife works better where large gaps need to be filled.

Polysulfide rubber is highly elastic and long-lived (more than 20 years) and can be used for moving joints and in areas exposed to extreme weathering, such as around skylights. It is more expensive than butyl but performs better.

Where to Apply Sealants

Four places on exterior walls require careful caulking, usually on a periodic basis.

First is between the frame of a window or door and the main frame of the house. The top of the head jamb is especially vulnerable, but the sides of a frame and the lower side of the sill should also be checked for any cracks.

Second is within the frame of a window or door. One especially vulnerable point is between the track of a sliding door or window and the sill on which it rests (or the jamb it hangs from).

Third is any gap between lengths of siding or where two different building materials meet. Shrinkage sometimes opens relatively wide gaps.

Fourth is any point where a deck or other flat surface abuts a wall or where any framing member protrudes through a wall. The joists that support a deck will need occasional caulking, as will ceiling joists that extend to form a patio roof.

WHAT ABOUT INSULATION?

Drafts and discomfort may be the first harbingers of winter if your home is under-insulated, even in mild climates. Adding insulation is a job you might do yourself, while working on walls or ceilings. And it may pay for itself in just a few seasons through lowered heating and cooling costs. To insulate properly you should consider two problems: heat loss and moisture control.

Household heat escapes through cold, uninsulated walls and ceilings and is replaced by cooler air, creating strong floor drafts, burdening your heating system, and leaving you chilled. Insulation retards outward heat movement by breaking up air space between walls into tiny pockets. It can provide more than four times the heat resistance of an uninsulated wall.

The moisture in escaping air condenses as it cools and can soak into the house framework and blister exterior paint. That ominous winter dripping may just be condensation, not a leak, but the damage can be just as great.

To provide a vapor barrier to stop the moisture flow, separate the insulation from the interior wall by a thin layer of foil, asphalt-impregnated kraft paper, or polyethylene film. Some forms of insulation are made with vapor facings. (Foil also reflects back what little radiant heat comes through the wall.)

Your ceiling is by far the most vulnerable to heat loss and may be the easiest to insulate if you have an accessible attic. Other important areas are outside walls, dormers, walls facing unheated rooms, and floors over cold basements or crawl spaces. You'll have to open up the walls to add insulation and should probably wait until you're ready for general remodeling.

Three main forms of insulation are available:

Flexible blankets in rolls or shorter batts, up to 6½ inches thick and made from mineral or glass fiber, are stapled between studs or joists and are easiest to use.

Loose fill for hand pouring is commonly either vermiculite or mineral wool. It may be the easiest way to insulate an attic or other above-ceiling space, but it is difficult to install fully into walls. Its cost is about the same as blanket insulation.

Many building supply stores carry a loose-fill cellulose insulation for use with a loan-out blower.

Rigid insulation boards are used most during construction, less for corrective insulation. One fiberglass panel has an attractive vinyl surface and can increase the heat resistance of an exposed-beam ceiling. It is considerably more expensive.

Some basic tips. A good average insulation thickness for attic floors is 4 inches in moderate climates and 6 inches in colder extremes; relative benefits diminish with greater thickness. The 3½-inch blankets are adequate for most floors and walls. You don't have to fill to the full depths of the joist or stud space. All that's needed is proper insulation thickness with an unbroken surface. For added moisture protection, fully vent all crawl spaces and attics.

You should discuss difficult insulation problems with a licensed insulation contractor. He can give a precise measure of your needs and potential savings, and he may be able to do the work more efficiently himself (his work should be guaranteed). Ask for a free estimate. You'll find that the cost of professionally installed ceiling insulation varies considerably. For more information, see the *Sunset* book INSULATION AND WEATHER-STRIPPING.

PRESERVATIVES AND SEALERS

The life expectancy of an unprotected board sunk in soil or subjected to heavy moisture is about the same as that of a gladiator in the ring with a hungry lion. Actually, the only real difference is the size of the enemy—lions are large; termites and fungi are on the small side. But the effect of both is equally devastating.

Termites and fungi, a persistent cause of decay, flourish in unprotected, damp woods. But wood preservatives, either soaked into wood or forced under pressure into wood fibers, create an environment that is hostile to decay. When properly applied, these modern preservatives make low-resistance woods, such as pine, cottonwood, and aspen, last as long or longer than the more naturally durable redwood and cedar species.

Decay is most likely to occur at points where wood touches or enters the ground, in areas of little air movement, and around joints between wood and other materials. Exterior decay often occurs around wooden steps and decking, along the bottom edge of siding, and around door and window sills. Inside, wood decays in places where moisture is generated by bathing, laundering, or dishwashing.

Several types of preservatives are listed below. Clear resin sealers, mentioned last, are in a category of their own.

Wood preservatives can be divided into two main classifications: water borne and oil borne. Among the water-borne are salt preservatives, known chemically as

Acid Copper Chromate (ACC), Ammoniacal Copper Arsenite (ACA), Chromated Copper Arsenate (CCA), Chromated Zinc Arsenate (CZA), and Fluor Chrome Phenol (FCAP). All of these salt preservatives are marketed under various trade names. Having several positive traits, they are highly recommended where clean, odorless, and paintable wood is required.

For greatest effectiveness, though, water-borne salt preservatives must be applied under pressure, a technique impossible for the home owner. Two other drawbacks: the salts are poisonous to plant life; and, because they are soluble in water, they leach out more readily than the oil-borne preservatives.

Best known in the oil-borne category is that granddaddy of preservatives, creosote. Two others popularly used are pentachlorophenol and copper naphthenate.

Creosote

Solutions of creosote have been successfully used for decades to protect telephone poles, railroad ties, fence posts, and pier foundations. Long lasting and insoluble in water, creosote is a very effective preservative, particularly for wood touching the ground. In residential construction, creosote is generally used only below ground level because it has a heavy odor and cannot be painted over in a normal manner (a brown stain will quickly seep through the paint).

Although lumber will last longer if pressure-treated, the most practical way for the homeowner to apply creosote is to soak it into the wood. Brushing creosote on wood to be buried in soil is unsatisfactory because creosote will not soak deeply into wood fibers.

Pentachlorophenol

As effective as creosote except for underground use, pentachlorophenol is clean, odorless, and leaves no stain if applied with a clear oil. Obtainable under several trade names, it comes in two forms. One, called a toxic wood preservative, is simply the chemical dissolved in oil. The other, a toxic water-repellent wood preservative, contains resins which make the wood nearly impervious to water penetration. This is a good choice for protecting girders that touch the ground, for deck posts and for sealing cross-bracing on posts exposed to weather.

Evaporation of the oils in pentachlorophenol's base leaves an elastic film on the wall of each cell reached by the preservative. Because of this, preservatives help to control cracking and swelling that comes with age.

You can buy commercially pressure-treated lumber or soak your own. Commercial treatment is more effective than soaking. Soaking time for home treatment of wood varies with wood species. For pine, 12 to 24 hours is enough. Redwood and cedar may need to soak as much as a week.

Avoid spilling pentachlorophenol on any plant material. Since the liquid has a secondary use as a weed killer, it will annihilate a broad-leafed plant on contact. It is also potentially dangerous to you; prevent it from contacting the skin as much as possible, especially keeping it away from your mouth and eyes.

Copper Naphthenate

Though more expensive than pentachlorophenol, copper naphthenate is favored by many homeowners because it is safe to use around plants. Though having a characteristic green tinge, it can be painted over. But don't apply it over an existing coat of paint or varnish. Soluble in water, copper naphthenate will leach out.

Clear Resin Sealers

Several available resin-based wood sealers can be brushed on or soaked into the wood. Not as durable as other preservatives, they are used mainly for sealing lumber against water penetration that would cause swelling and shrinking. Although they do not penetrate deeply into wood, they do preserve natural wood color and can be painted over. Soak lumber for 5 or 6 minutes in the solution, or splash the sealer on liberally with a brush.

Preservative Application Tips

For best results, wood preservatives should be applied under pressure. The effectiveness of the treatment depends upon the amount of preservative that is absorbed by the wood fibers. Much more fluid can be forced into the wood under pressure. Since this procedure requires large and expensive equipment, it is almost always done by a commercial firm.

The largest problem you'll have in soaking your own lumber is finding a tub that's large enough. The ends of short boards and posts can be set in a 50 gallon oil drum filled with preservative. For longer lumber, you can improvise, using two stacks of boards and a sheet of heavy polyethylene as illustrated.

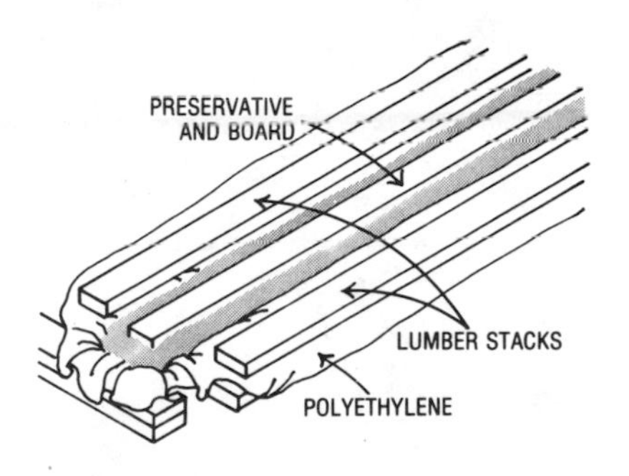

Some woods will take longer to soak up a preservative than others. Researchers report that pine will absorb the proper amount of pentachlorophenol for efficient preservation (4 to 5 pounds per cubic foot of wood) within 24 hours. Douglas fir, redwood, and incense cedar require longer periods, sometimes as much as a week.

If soaking is impossible, paint the wood with the preservative. Literally sop it on.

House Framing Index

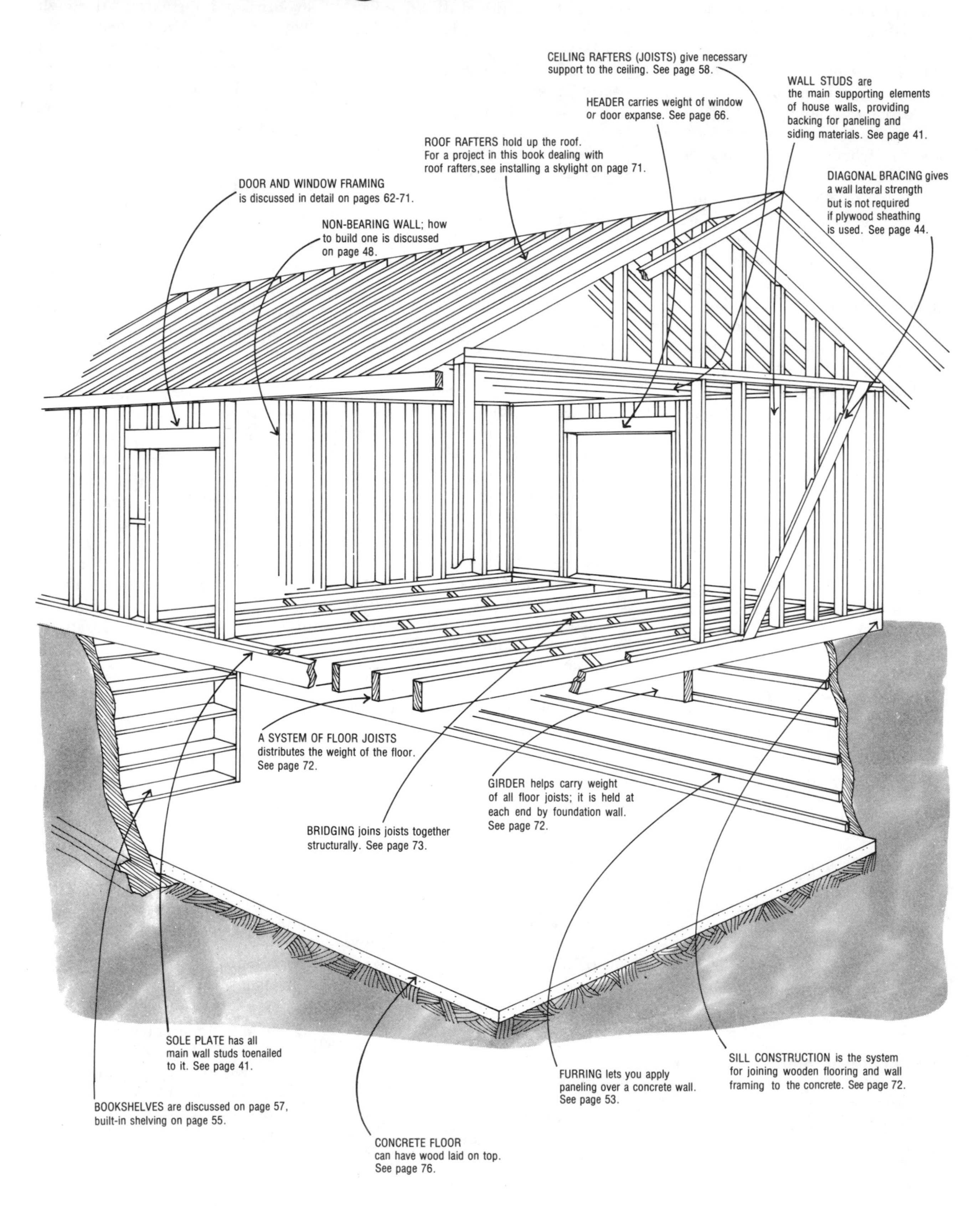

House Covering Index

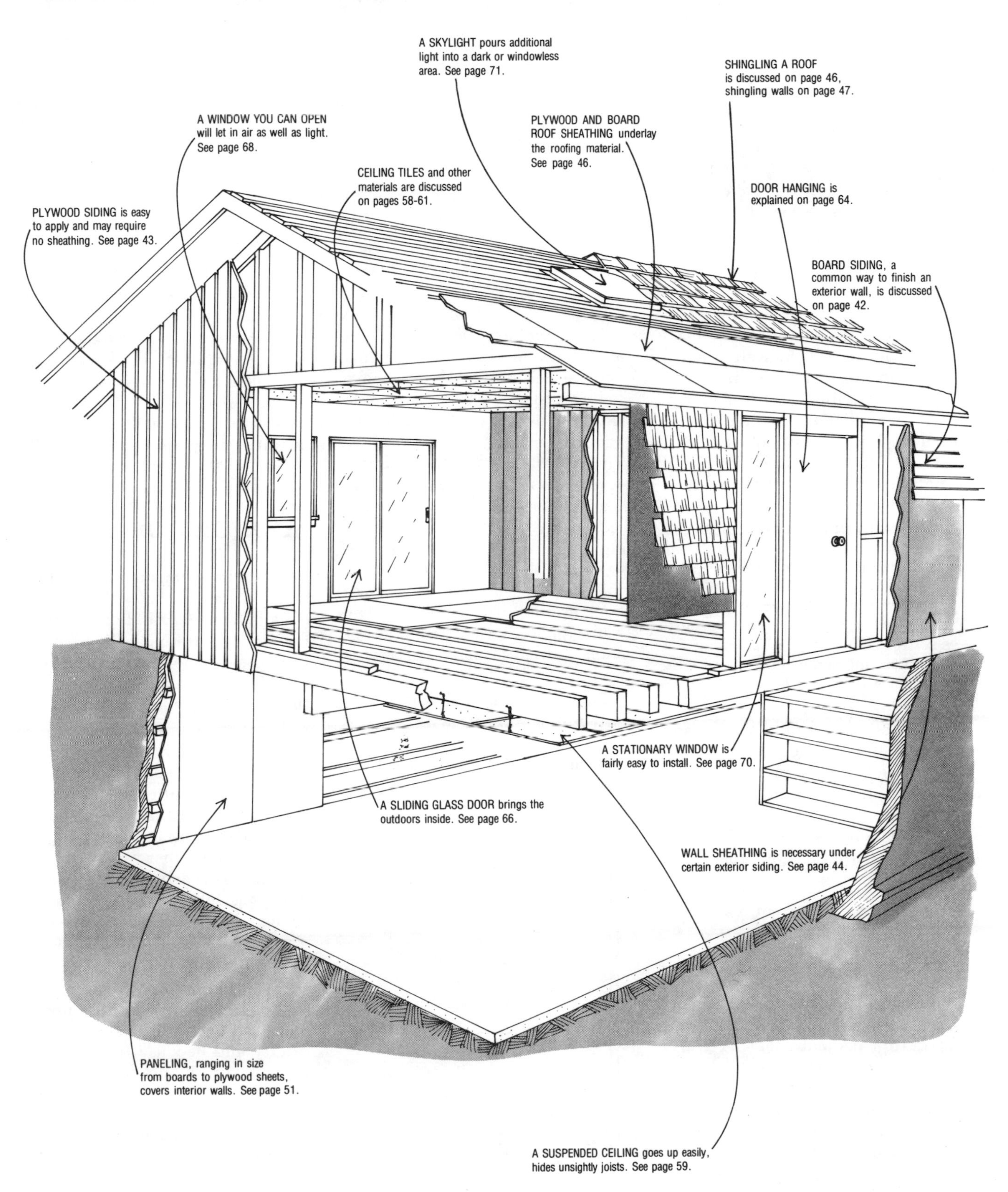

Walls

Carpenters make walls their business. They're hired for wall-oriented jobs ranging from building new walls to installing built-in shelving. That makes the subject of house walls a good place to begin a discussion of applying basic tool and material information to practical carpentry projects.

To give you an understanding of what you'll be working with, this section will first take a look at basic wall structure. Then we'll plunge into some projects—both large and small—that you can do yourself, saving the carpenter's fee for another time when you may really need professional help.

Gain familiarity with terms and specifications before attempting any project: if you're not informed, you can quickly reduce a two-story house to an oversized pile of lumber. In addition, check building codes to find out whether you will need a building permit for the project.

HOW TO DETERMINE A BEARING WALL

Knowing the difference between a bearing and non-bearing wall is crucial before beginning any wall construction project. Bearing walls carry part of the weight of the house; non-bearing walls do not.

All exterior walls that run perpendicular to ceiling and floor joists (and usually at least one main interior wall running parallel to these walls) are bearing.

To determine whether or not an inner house wall is bearing, climb into the attic or crawl space above ceiling joists. Check to see if joists are joined over any particular beam. The wall below such a beam will be a bearing wall. One note of caution: even if joists are long enough to span the house's length, their midsections may be resting on a bearing wall. Consult building codes to figure the allowable span of joists, and then measure these distances from exterior bearing walls. Any wall supporting the joists at maximum allowable span is a bearing wall.

Because of stress and load factors, only minor projects with bearing walls are covered in this book. If you wish to do major work on a bearing wall, consult a contractor or an architect.

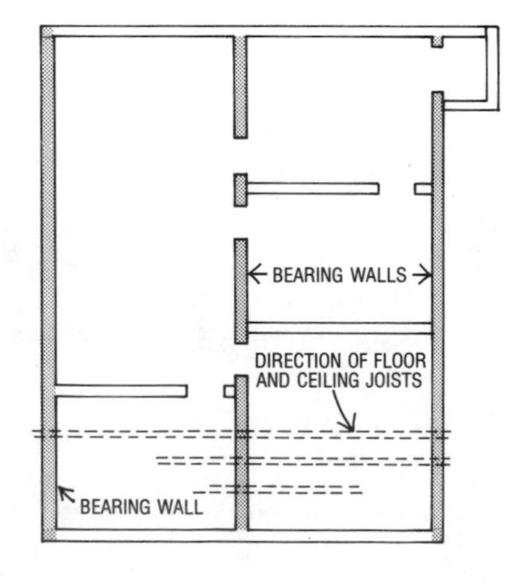

TYPICAL BEARING WALLS are shaded.

Other walls throughout the house serving simply to divide rooms are non-bearing walls. Lighter materials and less-precise specifications are often used for these. This is the type of wall chiefly discussed in BASIC CARPENTRY.

ANATOMY OF A WALL

Beneath the wall paneling or siding that you view as the "wall" is a systematic framework of lumber. Bearing and non-bearing walls have the same basic framework, but a bearing wall is more complete because its function is more important.

Lumber for wall framing is characterized by stiffness, nail-holding ability, and freedom from warp. Commonly-used wood species include Douglas fir, white fir, hemlocks, spruces, southern yellow pine, and other pines. The top two grades of each species are used, lumber is seasoned, and moisture content does not exceed 19 percent.

Two major systems of house framing are used in the United States: *western* (or *platform*) framing, and *balloon* framing. Western framing is found most often because of its simplicity and because this framing tends to shrink uniformly throughout the completed house. Though balloon framing also shrinks, it is generally used when masonry is to cover exterior walls because settlement at joists will not be as severe as with western framing.

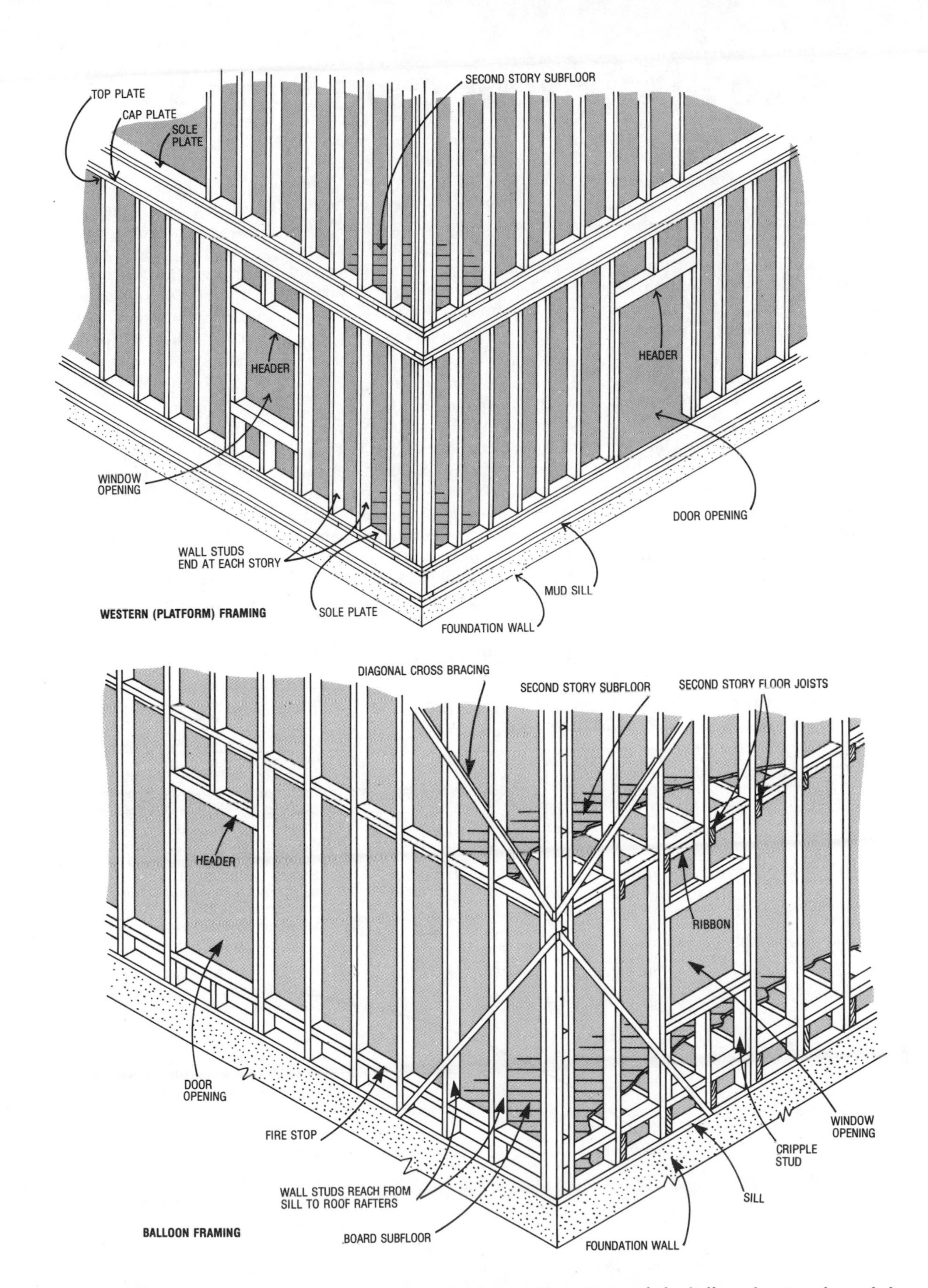

THE CHIEF DIFFERENCE between the western framing shown in the top illustration and the balloon framing shown below is that balloon framing has wall studs extending in one length from the foundation to the roof rafters two stories above. Western framing is more compartmentalized, with wall studs ending as a unit at every floor level.

How to Apply Your Own Siding

Has your house been presenting its tired old face to the world for too long? If it has been sagging, flaking, wrinkling, or just generally falling apart, the time may be ripe for drastic steps. If you don't think cosmetics will adequately cover up the damage, you'd probably better consider a complete face lift. In the case of a house, this means applying new siding.

Not only old houses need siding; so do new houses. If you're having one built or remodeled, you might be able to save considerable money by arranging to apply the siding materials yourself.

What will this entail? Your work would vary, depending upon the kind of siding you choose. It is usually a simple matter of nailing siding materials to the existing frame or surface. Some sidings require a backing of wall sheathing (page 44) when applied directly over studs; others don't. Application varies even within the category of board siding, but the main difference occurs between board siding and sheet plywood siding: plywood goes up much faster.

Both plywood and board sidings are discussed in full below. For information on shingling house walls, see page 45. Manufactured sidings generally requiring special skills or techniques apart from carpentry—aluminum, vinyl, pressed hardboard, mineral fiber, and asphalt—are not discussed. Get in touch with the manufacturers of those products for information. Before beginning work, don't fail to check the building codes in your area to see if a building permit is required.

Before applying siding, a strong nailing base is necessary. You may need to attach a grid of 1 by 3 or 1 by 4 nailing strips. Patterns of grid application will vary depending upon the siding material, but fastening methods are basically the same. Attach these strips with nails that are long enough to penetrate at least 1 inch into wall studs.

Use only high-tensile strength rustproof nails (see page 29.).

SOLID-BOARD SIDING

Solid boards have been around houses since the early days of log cabins. Unlike the rough logs used then, though, today's solid-board

APPLYING BOARD SIDING

SIDING TYPE	DIRECTION OF APPLICATION	NAIL SIZE	NAILING TIPS
Board on Board (unmilled)	Vertical	8d underboards 10d overboards	Face-nail underboards once per bearing, face-nail overboards twice. Minimum overlap 1″. Nailing blocks are necessary between studs.
Board and Batten (unmilled)	Vertical	8d underboards 8d or 10d battens	Space underboards ½″ apart. Face-nail each board once per bearing. Minimum overlap 1″. Nailing blocks are necessary between studs.
Clapboard (unmilled)	Horizontal	10d	Face-nail 1″ from overlapping edge (just above preceding course) once per bearing. Minimum overlap 1″.
Bevel	Horizontal	8d for ¾″, 6d for thinner	Face-nail once per bearing. With rabbeted bevel, face-nail 1″ from lower edge. Allow expansion clearance of ⅛″. Minimum overlap 1″.
Shiplap	Horizontal or vertical	8d for 1″, 6d for thinner	Face-nail once per bearing for 6″ widths, twice (about 1″ from overlapping edges) for wider styles. Nailing blocks over open studs are necessary for vertical application.
Tongue-and-Groove	Vertical, horizontal, or diagonal Can mix widths	8d (*finish* for blind-nailing, otherwise *siding nails*)	Blind-nail 4 to 6-inch widths through tongue with finish nails, once per bearing. Face-nail wider boards with two siding nails per bearing. Nailing blocks over studs are needed for vertical application.

sidings are professionally produced finish materials.

Although sometimes lengths of standard board lumber, such as 1 by 8's or 1 by 10's, are used for siding, boards are usually milled into special siding patterns that have interlocking grooves, lapping edges, or bevels. As a rule, the amount, type, and grade of lumber needed and the number of cuts required to produce board siding determine its cost.

Siding woods should be good for painting, easy to work with, and free from warp. Because they have these traits and because of their natural resistance to decay, redwood and cedar heartwoods make excellent solid-board sidings. Other good woods include eastern white pine, western white pine, sugar pine, and cypress. Fairly good woods for siding include western hemlock, ponderosa pine, spruce, and yellow poplar. Species that are only fair for sidings are Douglas fir, western larch, and southern yellow pine.

No matter what type of wood you choose, pick "Clear" or "Select" vertical-grain boards that are primarily of heartwood because vertical-grain surfaces won't warp as easily as flat-grain. Tight knots in the boards are acceptable if you treat them before applying a finish. Be sure the wood is always "Certified Kiln Dried" and stamped with a known trademark. For more information on lumber in general, see page 24.

For your convenience, some solid-board sidings can be purchased preprimed with paint or pretreated with water repellent or a sealer.

Although most board sidings are applied either horizontally or vertically, some (like tongue-and-groove) can be placed vertically, horizontally, or even diagonally.

Board sidings that lock together (tongue-and-groove or shiplap) generally provide the most protection and are easier to apply because they automatically align with each other.

Figure siding course locations so that boards above and below windows and above doors can run continuously without notching. Be sure to butt boards snugly together at corners and at doors and windows.

Once board siding is delivered to the building site, handle it carefully. Because most siding woods are quite soft, they are easily damaged. Protect boards from moisture by keeping them off the ground and under cover.

Board sidings will probably split when nails are driven near the edge unless the nails are blunted or you predrill nail holes. You can buy nails blunted or blunt them yourself with the hammer. Try not to split the sidings.

On the facing page is a chart of the primary board siding patterns and their application specifications. These patterns are basic but many decorative variations are available.

PLYWOOD SIDING

Plywood's large panel size (4 by 8, 9, or 10 feet), its strong laminated construction, and its variety of surface styles make it a popular siding. Not only does it lay up rapidly and easily but also it can give a wall extra rigidity and eliminate the need for crossbracing and wall sheathing on the wall's frame.

Many textured varieties of exterior plywood, such as roughsawn, striated, and grooved, are used for siding as well as for other construction purposes. For general information about plywood types and how to work with plywood, see page 25.

One plywood product made specifically for siding—horizontal-lapped plywood siding—is not purchased in standard sheet form but in long panels of 12 to 16-inch widths and 16-foot lengths.

Whatever pattern or style you choose, be sure to specify *exterior-grade* plywood.

Plywood siding applied directly to studs without sheathing must be at least ⅜-inch thick for studs on 16-inch centers and at least ½-inch thick for studs on 24-inch centers. Thicknesses as narrow as 5/16- inch may be applied over wall sheathing or firm older walls.

Applying Plywood Siding

Large plywood panels can be applied either vertically or horizontally. If you apply them over studs horizontally, stagger the vertical edge joints and nail the long, horizontal edges (that join four feet up from the base) into firestops or other nailing blocks. Keep panel edges on center of studs and nailing members. Nail every 6 inches along edges and every 12 inches on intermediate supports. Nail sizes will differ with panel thickness. Use 6d nails for ⅜-inch or ½-inch panels and 8d nails for ⅝-inch panels. For lap or bevel sidings that are ½-inch or thicker, use 8d nails.

Like any siding, plywood must

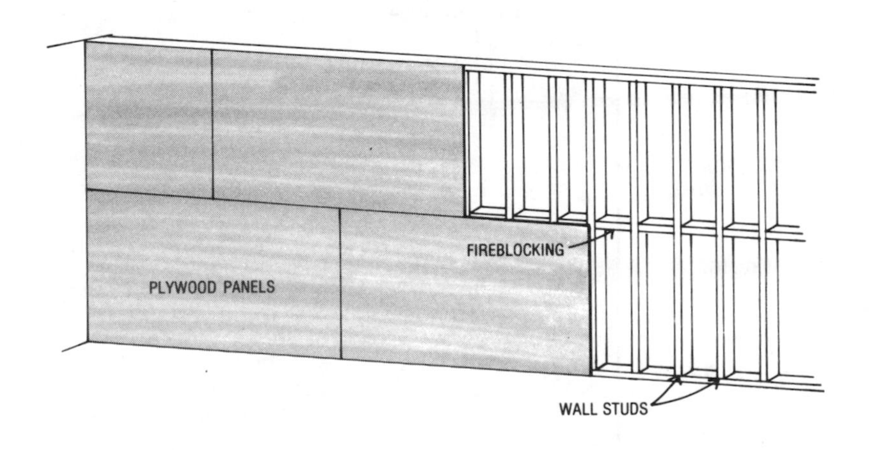

keep moisture and weather out. Seal all edges with a prime coat of paint or a water-repellent preservative. Leave 1/16-inch space between all panel ends and edge joints for expansion under moist conditions. See page 35 for information on caulking and flashing.

WHEN TO SHEATH

Though many siding materials give a wall sufficient strength and rigidity, sheathing is needed beneath others for rigidity, insulation, and sometimes to serve as a solid base for nailing.

Sheathing is generally only applied to new structures. The previous siding of older walls usually acts as sheathing for new coverings.

Several types of sheathing are commonly used: plywood (the most common), exterior fiberboard, exterior gypsumboard, and regular board lumber. Plywood is most frequently used because its large-sized panels simplify application and may afford enough lateral strength to eliminate the need for diagonal wall bracing in the framing.

Below is a chart comparing the main types of sheathing and basic application techniques. All types should be nailed directly into studs. As a rule, use nails long enough to penetrate studs at least 1 inch. Use waterproof spiral nails unless otherwise specified (see page 29).

WALL SHEATHINGS: HOW THEY RATE

Qualities	TYPES			
	Ext. Plywood	**Ext. Fiberboard**	**Ext. Gypsumboard**	**Solid Board**
Direction of application	Vertical or horizontal	Horizontal	Vertical or horizontal	Diagonal or horizontal
Panel sizes and types	5/16, 3/8, 1/2-inch thicknesses in panels of 4 by 8, 9, or 10 feet. Square-edge or tongue-and-groove.	1/2, 25/32-inch widths in 2 by 8-foot panels. Tongue-and groove or shiplap.	1/2-inch widths, 2 by 8-foot panels. Tongue-and-groove.	1 by 6: end-matched tongue-and-groove. 1 by 8 or 1 by 12: shiplap.
Rigidity	Good	Fair	Good	Good
Insulative value	Low	Good	Low	Fair
Nailing	Nail every 6 inches along panel's edge and every 12 inches into center supports. Heavy-duty staples spaced closer may be acceptable.	Use roofing nails 3 inches apart along edges, 6 inches apart intermediately.	Gypsum nails every 4 inches around edges and every 8 inches intermediately.	Three nails per bearing for widths of 8 inches or more, 2 per bearing for lesser widths.
Does wall need diagonal bracing?	No	Yes, with standard types	In some areas	Only for horizontal application
General notes	Use STANDARD grade. Apply with panel ends spaced 1/16-inch apart and edges 1/8-inch apart.	Easy to handle and apply. Don't nail within 5/8-inch of edges. Only a special type will serve as sole nailing base for siding.	Not a nailing base for siding.	Also available are 2 by 8-foot panels made from edge-glued lumber, overlaid with building paper.

Using Shingles and Shakes

For those who are quick to answer the call of the great outdoors, shingles and shakes may be just the ticket. Their rugged, woody appearance provides a rustic flavor that no other building material can match.

Shingles and shakes can cover both roofs and walls of houses. Sometimes they're even applied to interior walls (see page 51). Aside from their handsome appearance, they admirably defend a house's exterior from onslaughts of rain, wind, and snow.

The location of shingles determines the job they'll have to do. On the roof they probably will have to battle severe precipitation and the effects of heat. On exterior walls they'll have to repell moisture and cold but they won't have to work nearly as hard. Applied as decoration to inside walls (page 51), shingles or shakes will get much less wear.

Types of Shingles

Shingles vary according to grades, sizes, textures, and cuts. Quality is graded by numbers ranging from 1 to 4. Number 1 shingles, made from 100 percent vertical-grain heartwood, are the best. Number 2 is a good grade of shingle that allows minimal flat grain and sapwood. Number 3 is a utility grade, and Number 4 is for under-coursing and use on interior walls.

You are expected to specify shingle lengths, but widths are random unless you ask for "dimension shingles." Common lengths are 16, 18, and 24 inches. Widths vary from 3 to 14 inches.

To greatly reduce labor time of laying shingles, you can buy large panels of shingles bonded to a backing of ½-inch sheathing-grade plywood. Typically, these panels are long and thin—18-inch shingles placed on 8-foot strips of plywood.

For shingling sidewall, specify "rebutted-and-rejointed" shingles having all edges trimmed square. For decorative purposes, purchase dimension shingles with butts especially trimmed to half-round, diamond, half-cove, acorn, or other unusual shapes. They are also available prepainted or prestained.

Types of Shakes

Shakes are like large, thick hand-split shingles. Three different variations of shakes are available: hand split and resawn, taper split, and straight split.

Hand split and resawn shakes are thin at one end, thick at the other; rough on the face, smooth on the back side.

Taper split shakes are also thick at one end and thin at the other, but because they are not resawn, both faces are rough.

Straight split shakes have a regular thickness and are rough on both sides.

All shakes are Number 1 grade. Thicknesses vary from ⅜-inch to 1¼ inches; lengths are 15, 18, 24, and 32 inches. Widths are random, and not all types of shakes are available in all sizes. Check with your dealer about sizes available for the type you want. Chances are, he'll give you a fair shake.

Some Shingling Terms

Before he receives information about using building paper and sheathing or reads about techniques for applying shingles and shakes, a carpenter needs to expand his shingling vocabulary.

Here is a list of terms commonly used in shingling:

Pitch is the degree of slope a roof has; it determines if and how shingles or shakes should be applied. Pitch is measured by the number of inches a roof rises in a level distance. For example, "4 in 12", means the roof rises 4 inches for every 12 inches it runs.

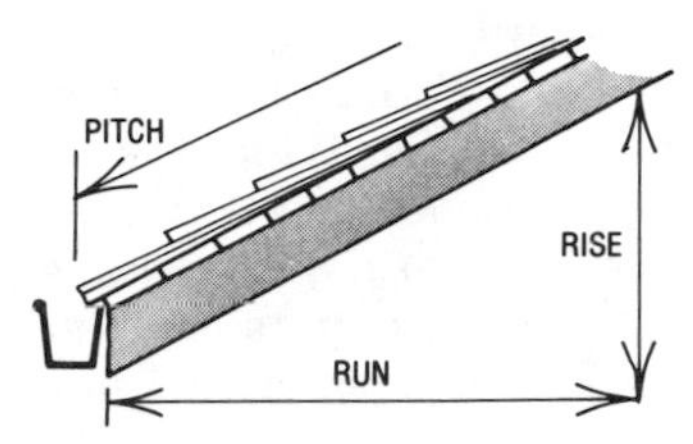

Course is a row of shingles or shakes placed side by side (actually spaced ⅛ to ¼-inch) along the length of the roof. Each course overlaps two-thirds of the course directly below it and one-third of the course below that. This insures having three layers of wood at all points on the roof. The bottom course, along the eve line, is always doubled for added thickness.

Exposure signifies the part of a shingle or shake that is open to the weather. The amount of allowable exposure depends on the steepness of roof pitch.

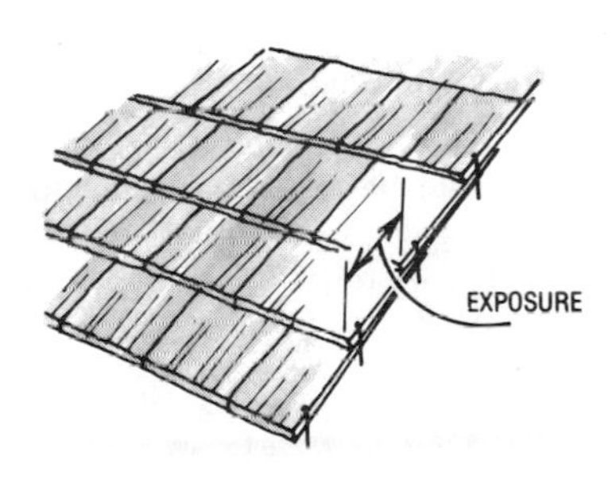

Maximum exposure increases with steepness. On roof slopes of 4 in 12 or steeper, standard exposures for Number 1 grade shingles are 5 inches for 16-inch shingles; 5½ inches for 18-inch shingles; 7½ inches for 24-inch shingles. Reduced exposures are recommended on roof slopes of 3 in 12 up to 4 in 12 and for all slopes for Number 2 and Number 3 grades.

Standard exposure for 18-inch shakes is 7½ inches; for 24-inch shakes, 10 inches.

Square is the quantity measurement for shingles and shakes. One square is the amount needed to cover 100 square feet of surface at standard pitch and exposure. Depending upon the size and shape of shingles or shakes, one square may be 2, 4,

5, or 6 bundles; for most shingles, one square is 4 bundles.

Hips, valleys, and ridges are the different angles formed by the planes of a roof. The ridge is the peak along the top, hips are other ridges at the junctures of two sloping sides, and valleys are the gutters (or angles) formed by two meeting roof slopes.

Sheathing and Building Paper?

Typical roofs have either solid sheathing or open-board sheathing. Solid sheathing is often preferred because it is more structural.

Shingles offer a surface that's practically impervious to water, but it is a good idea to staple a layer of 30 lb. building felt over the sheathing before shingling. Layers of building paper should overlap by at least one third their width.

Unlike shingles, shakes have a tendency to let a certain amount of moisture blow through in severe climates. In applying shakes, place strips of 36-inch-wide 30-pound roofing felt over solid sheathing at the eve line, lapping the top ends of each course as shown. Where the climate is very mild, open sheathing may be used.

Application to Roofs

If there's something worse than a leaky roof, most homeowners haven't come up against it. Dripping water can puddle up, ruining ceilings, rugs, and furniture. Because finding leaks in established roofs can be a chore, it's important to do the job right in the first place.

Apply shingles and shakes as shown in the two sketches at right, beginning with a double course at the eve line and working upward. The outer edge of the first double course should protrude slightly over the gutter's inside lip so that runoff water will spill away from the eve into the gutter.

To keep courses properly aligned as you work, snap a chalk line across each course where butts of the next course should align. Temporarily tacking a long straight board along this line will allow you to rapidly align shingle butts as you nail.

Nail each shingle or shake with two rust resistant shingling nails (page 29). Space the nails less than ¾-inch from side edges of shingles, 1-inch from side edges of shakes. Shingle or shake sizes will determine the length of nails to use; they should penetrate well into sheathing. For 16 or 18-inch shingles, use 3d shingling nails. For 24-inch shingles, use 4d. Most shakes can be nailed with 6d shingling nails, but thick shakes or severe weather require longer sizes. Don't drive nailheads in such a way that they smash the wood.

Hips and ridges cap the corners where the roof forms a crest. You can either buy premade hip and ridge units or make your own by lacing uniform-width shingles together as shown. Nail with two 8d

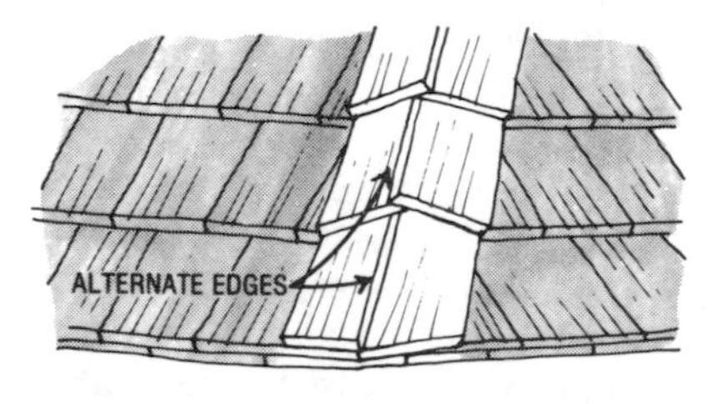

shingling nails on each side of each unit, penetrating the sheathing.

Where the roof forms valleys, you must apply special valley flashings to collect runoff water and send it to the gutters. Use metals proven reliable in your region. The sheets should be at least 20 inches wide and crimped along their centers. For shakes, apply 30-pound roofing felt beneath flashings. Valley edges should extend well under the edges of each course. Cut shingles or shakes to the proper miter. Avoid laying them with grain parallel to valley. See page 34.

Apply flashings in the same fashion as for valleys wherever angles are formed on a roof, such as the angles where a chimney or dormer meets the roof.

Old shingled roofs can be covered directly with new shingles or shakes, saving you the labor and mess of stripping off all the old shingles. Doing this will give the house double insulation and protection, but such layering of shingles will not be possible if an old roof is extremely moist, rotting, insect infested or unable to support the load.

Old shingles should be cleaned of any debris, fungus, or moss. Remove any badly warped or buckled shingles, replacing missing shingles with new ones.

Cut away a 6-inch width of shingles along the roof's eves, ridges, and ends. Nail strips of 1 by 6 lumber in place of removed shingles along the eves and gable edges. Then nail a strip of cedar bevel siding along the ridges. In addition, nail a length of 1 by 3 redwood or cedar in each valley to separate the old flashing from the new. Check all other flashings, making sure they will extend sufficiently above the new roof's surface. Lay a new angle flashing along the lip of the roof edges above gutters as shown. Once these preparations are made, shingle or shake just as you would over sheathing.

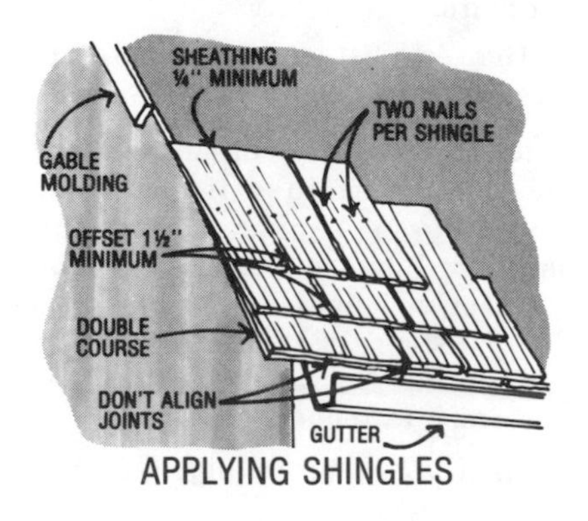

APPLYING SHINGLES

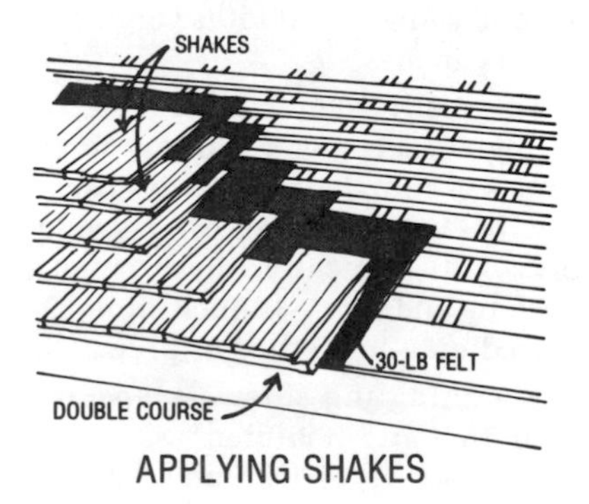

APPLYING SHAKES

Application to Walls

Shingles and shakes as exterior coverings have spanned generations. They have appeared on walls of houses varying in vintage from turn-of-the-century Victorians to futuristic domes. No doubt their popularity as a wall covering stems from their strong texture, toughness, and natural appearance.

Both shingles and shakes can be applied either in single or double courses. In single course application, each course overlaps the one below by at least one-half the shingle or shake's length.

Double coursing calls for two complete layers: an underlay of low-quality shingles (sometimes shakes), covered by a layer of high-quality shingles or shakes. Maximum exposure of finish shingles or shakes is achieved with this method.

To determine the number of courses and the proper shingle or shake length and exposure, measure a main wall from top to foundation. Subtract about 1 inch for the area directly above the foundation, figuring the number of courses according to the table below:

COURSE WIDTHS
(Exposure)

	Lengths	Single Courses	Double Courses
SHINGLES	16″	7½″	12″
	18″	8½″	14″
	24″	11½″	16″
SHAKES	18″	8½″	14″
	24″	11½″	20″
	36″	15″	—

Next, mark the locations of the courses on a long, thin board, called a "story pole." Plan to vary courses slightly if necessary to keep shingle butts aligned with window sashes and eves.

After transferring the story pole marks to the building's corners and window and door casings, snap a chalk line along the lowest marks just above the foundation. Then, as you later attach shingles or shakes in courses from bottom to top, snap a chalk line at each successive mark so that it leaves a line across the preceding course.

Staple or nail 30-pound roofing felt along studs, sheathing, or the wall surface before beginning to apply shingles.

You can nail shingles directly to old wooden walls. To shingle over stucco, apply 1 by 4 nailing strips, spacing the centers of the strips 2 inches above the butt line of each shingle course.

To nail shingles in single courses, use rust-resistant 4d shingling nails (see page 29). Begin with a double course (two thicknesses) at the bottom, nailing about 1 inch above the chalk line where the butts of the next course will align. Because the nails won't show, this is called "blind nailing." For shingles 8 inches wide and narrower, drive in two nails, each about ¾-inch from opposite edges. For wider shingles, drive a third nail in the middle. Set the nails snugly with the hammer, being careful not to crush the wood.

For the remaining courses, tack a long, straight board along the chalk line to serve as a guide for rapidly and accurately butting and nailing shingles into place. Before nailing each shingle, be sure the vertical joint between it and its neighbor is snug and that this joint does not align with a vertical joint in the course directly below.

Nail 1½-inch-square wood strips to any inside corners of the wall for shingles to butt against. Shingles at outside corners can either be mitered together or overlapped as

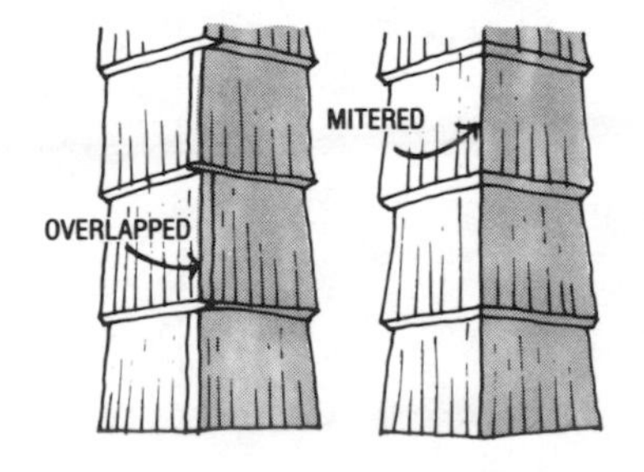

shown. If beveled corners do not fit true, shave them with a knife or block plane.

To fasten shakes in single courses, nail slightly above the next course's

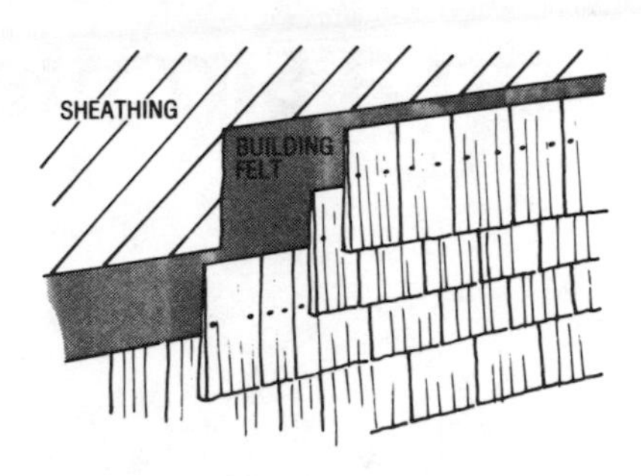

butt line and about 1 inch from each side of each shake. Nails must penetrate backing at least 1 inch. Otherwise, follow the same directions as for single-coursing shingles.

To nail shingles in double courses, use rust-resistant 9d shingling nails.

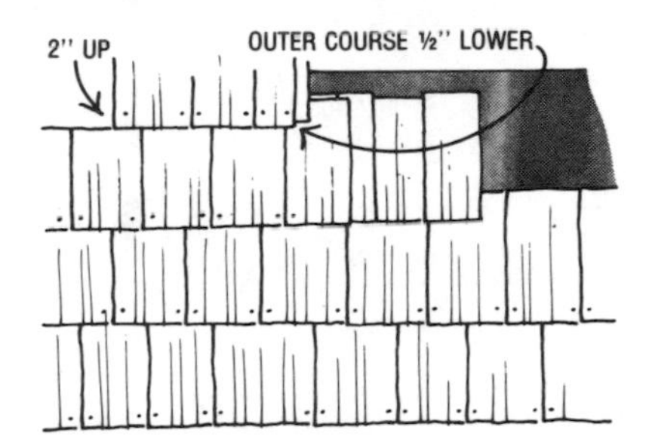

The bottom course should be made up of three separate layers of shingles: one extra undercourse on top of the regular undercourse, and the finish course. Nail or staple each undercourse shingle once; spacing their vertical edges about ⅛-inch apart. After stapling or nailing both thicknesses of undercourse shingles across the bottom course, fasten each finish shingle on top (but about ½-inch lower) by nailing about 2 inches above that shingle's butt, ¾-inch in from each side. Add additional nails spaced approximately 4 inches across the shingle's face. Use 5d small-head rust-resistant nails.

Vertical joints between these finish shingles should be snug and not aligned with the vertical joints between the undercourses.

Using the chalk line and board for a guide, work up the wall in the same manner as that described for single coursing.

To double-course with shakes, nail shingles through the butts, as well as above the butt-line of the next course.

Add a Simple Wall for New Room

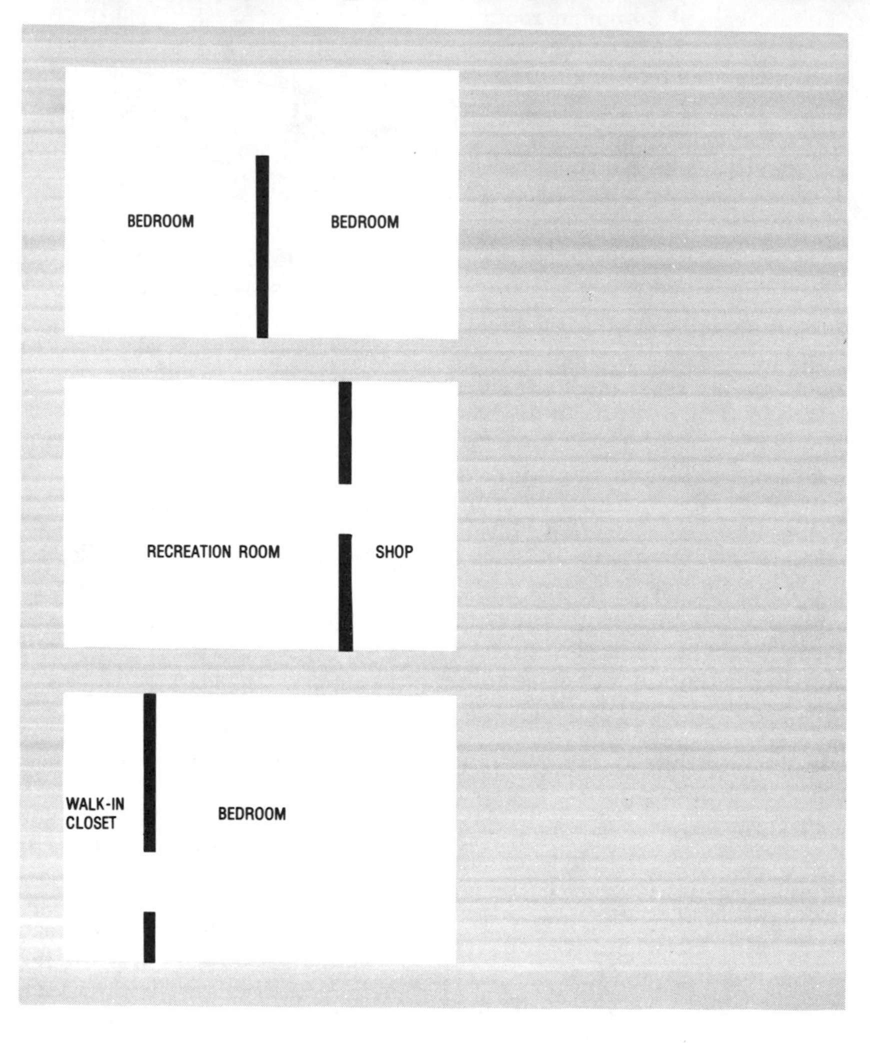

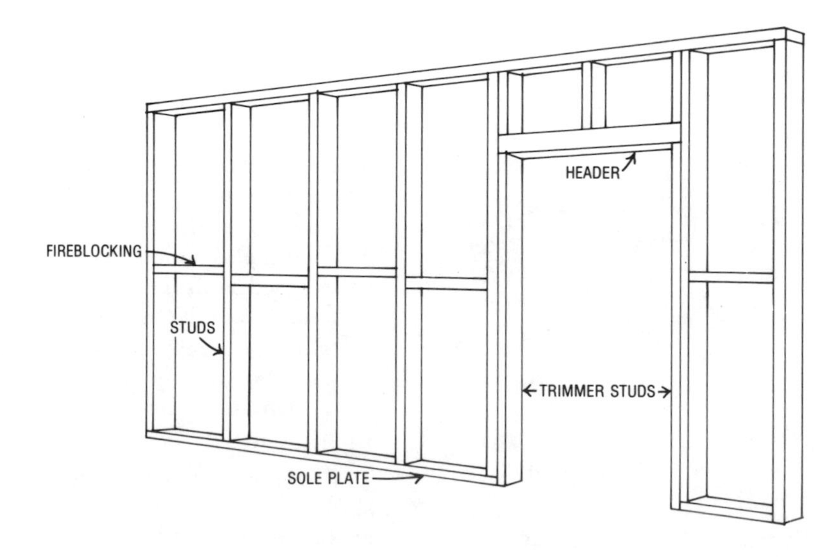

How long have you thought about that wasted space in your oversized recreation room—space that could be walled in to form a useful workshop or photo darkroom? Or are you more concerned with the need for separating the boys' room into semi-private living quarters? Whatever your project of space separation may be, the key to it lies in the simple matter of adding a non-bearing wall.

Although the thought of adding a non-bearing wall might tempt you to leaf through the yellow pages to find a professional carpenter, you'll find the job can be done quite easily if you plan properly and follow a few simple guidelines. Keep in mind that the information given here concerns non-bearing, *not* bearing walls. If you are unsure of some terms or specifications, refer to page 40.

When planning for a wall, remember that it will change the nature of the room, including lighting, heating, and traffic patterns. So it may be necessary to make additions or changes in wiring, plumbing, or heating during construction. If planning for some or all of these changes is beyond the scope of your abilities, consult a professional.

Framing for any non-bearing wall includes a sole plate on the floor, evenly-spaced studs, and a top plate along the ceiling. Firestops are almost always required. For walls with doorways or window-type openings, you'll need extra trimmer studs and headers.

Locating the Wall

Anchoring the new wall securely is most important. Placement of the wall will depend upon direction and location of ceiling joists above it and the position of existing wall studs at either end. The wall's top plate must be secured firmly to joists or firm nailing blocks. At least one of its ends should be nailed to an existing wall stud or nailing blocks. If walls are concrete, you can use special fasteners (see page 31).

Locate joists above ceiling material, following the approach for locating wall studs described on page 54. Or go into the attic or crawl-space above joists and drive a small nail down right next to each end of a joist. These nails will penetrate through the ceiling material and give you a handy reference point for measuring off the spacing of other joists (see information on ceiling framing, page 58.)

If your new wall will run parallel to joists and you can't gain access to the joist area, you *must* locate the wall directly beneath one joist. This way, the wall's top plate can be nailed to that joist.

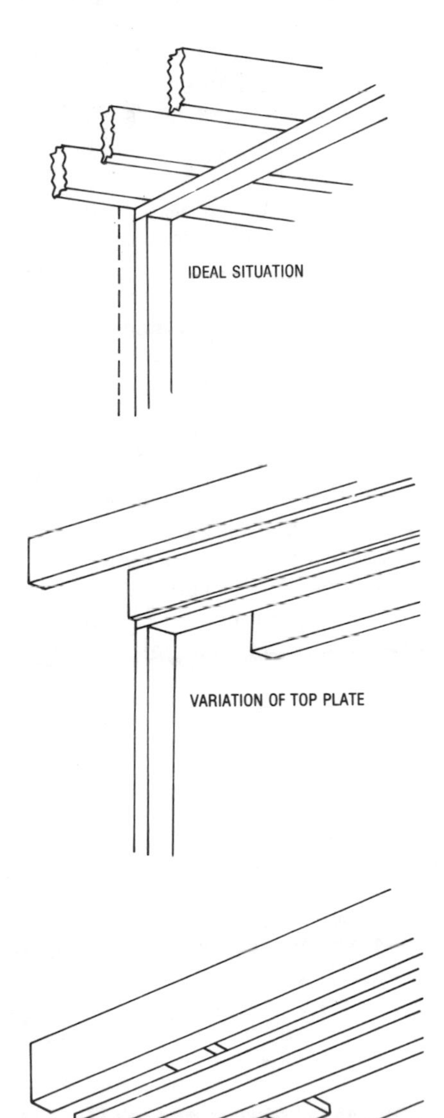

Try to determine the centerline of the joist and mark it on the ceiling at each end of the planned wall's location. Measure 1¾ inches to one side of each mark (half the distance of a 2 by 4 plate) and make second marks. Snap a chalk line between these second two marks. One edge of the top plate will run directly along this line.

If you have access to joists that run parallel to the proposed wall and don't want to locate it directly below one joist, use 16d nails to fasten 2 by 4 nailing blocks between the joists as shown. The top plate will later be nailed to these blocks. If you plan to install a ceiling, leave enough distance between these blocks and the bottoms of joists to attach a ceiling-backing strip flush with joist bottoms (see sketch). You'll need this for nailing the edges of the ceiling material along the wall's top, unless you add a suspended ceiling or ceiling tiles to nailing strips.

After marking the centerline for the top plate at each end of the planned wall's location, follow the procedures above for establishing a chalk line.

If joists run perpendicular to the proposed wall, you can freely locate the wall. Plan it where its end studs can be nailed to studs of the established walls. Follow the same procedures as above for establishing a chalk line.

Placing the Sole Plate

Hang a plumb bob from two points, one at each end of the top-plate line. Mark the floor; then snap a chalk line on the floor between the two marks.

Cut both the sole plate and top plate to length. Then lay the sole plate along its chalk line, positioning it so that it is directly beneath the top plate. Be sure to use both chalk lines to locate corresponding edges of the two plates.

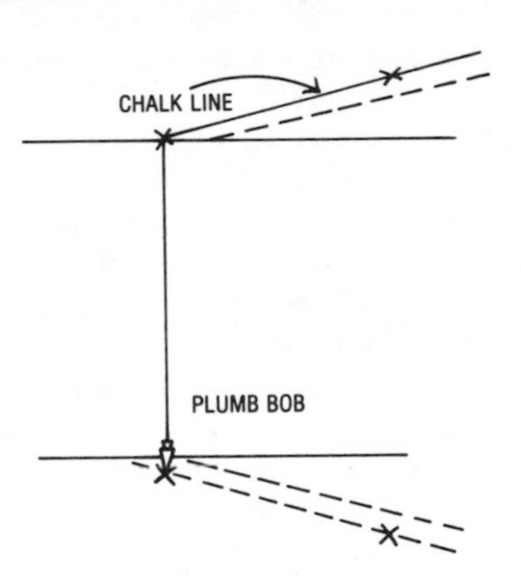

The type of flooring material that the wall will stand on will determine the method for fastening down the sole plate. For concrete, linoleum over concrete, or other masonry floors, drill two holes side by side into the concrete every 2 or 3 feet, using a masonry bit (page 17). Then insert metal or fiber anchors for lag screws (page 32).

Nail the plate to wooden floors, using 10d nails at 2-foot intervals. Be sure no heating ducts, plumbing, or wiring are below the floor where you are nailing.

Don't secure the plate to any floor within a section where a doorway will stand in the new wall because you will want to cut this section out later.

Marking for the Studs

Before securing the top plate, stack it on top of the sole plate, simultaneously marking the location of each regular stud on the edges of both as described below.

Beginning at an end that connects with an existing wall, measure in ¾ inch from the ends of the plates and make a mark. Using a combination square, extend this mark across the narrow dimension of both plates. Then starting from the wall, measure off 16-inch intervals to the other end of the plates. Each line designates the placement of a stud, the center of the stud sitting directly on the mark.

Although 16-inch spacings are standard, 20 or 24-inch intervals may be used when wall-paneling will fit these alternate spacings.

The last stud must be placed to provide an anchor. The marks for its location will depend upon whether your wall ends freestanding, turns a corner, or connects onto the opposite room wall. If the other end will be freestanding or connect with an existing wall, mark plates ¾-inch in from the ends, regardless of where the last regular 16-inch-spacing line falls. For a wall that will turn a corner, mark for stud placement according to the corner construction shown below.

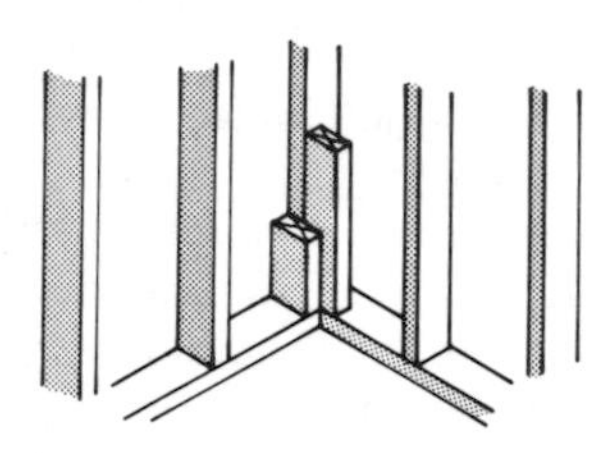

Next, mark the centerline of any door location or other opening.

In each direction, measure out one half the opening's distance, plus the jamb's thickness, plus ½-inch for shimming plus ¾-inch to overlap trimmer studs (the main supports of the door or window frame)—making X marks. These are the center marks for trimmer studs. On the other side of these lines, measure an additional 1½-inches and mark centerlines for full-lengthed studs.

For a window-type opening, mark off an additional ¾-inch for cripple studs inside each trimmer, above and below the opening. See additional information about framing doors and windows on page 62.

Fastening the Top Plate

Once both plates are marked, nail the top plate through the ceiling material or directly into joists. Nails should penetrate at least 1 inch into these nailing members (see page 49).

Putting It Together

Construct an outside corner post or end stud assembly as shown. Beginning at the end that connects to an existing wall, nail the new wall's end stud to the wall and toenail to top and sole plates. Use 8d nails. You may require special fasteners (page 31) where the wall does not meet an existing stud.

The *middle* of the first stud's narrow dimension should center directly on the marks ¾-inch from the wall. Cut each full-length stud to size, and toenail it into place with three 8d nails on the next set of marks at both the top plate and sole plate. If necessary as you work, mark a centerline at studs' ends to be matched with the centerlines on the plates.

A carpenter's trick: make a spacer-block 14½ inches long to fit between regularly-spaced studs as you toenail them in place. Such a block serves as a good brace for nailing. If you use one, make sure studs are on marks before nailing.

For doors or other openings (see pages 62-71), nail rough sills and headers into place after putting in trimmers and before adding cripple studs. Cut the sole plate out of the doorway section before adding trimmers. Extend trimmer all the way to the floor. Nail each trimmer to end of its sole plate and up its length to the full-length stud that it rests against. Stagger 16d nails back and forth across the trimmer's width.

Some building codes require an 8-foot wall to have horizontal bracing between the studs. The number of rows required depends upon local codes and the type of paneling material being used. If you plan to panel the wall with horizontal board or sheet paneling, probably only one row of bracing centered 4 feet up from the floor will be necessary. For vertical-board paneling, though, you'll need three rows equally-spaced between floor and ceiling to serve as a nailing base.

After all framing is completed, you'll want to consider inner-wall fixtures (plumbing and wiring), insulation (see page 36), and wall paneling (pages 51-54).

WHAT YOU NEED FOR A 20-FOOT WALL

Here is a list of the lumber requirements for a 20-foot wall in a room with a standard 8-foot ceiling. This sample wall has one doorway but no other openings or built-ins.

•**Twenty feet of straight 2 by 4** (redwood, if laid on concrete) for the sole plate. You can get either one 20-footer or two shorter lengths that total 20 feet. If you get two lengths, plan to join them either with in the doorway or more than 4 feet away from it.

•**Twenty feet of straight 2 by 4** for the top plate. Choose either one 20-foot length or two 10-foot lengths.

•**Fifteen regular 8-foot 2 by 4 studs** to be toenailed on 16-inch centers across the wall's length (except in the doorway) and face nailed at each end.

•**Two regular full-length studs** not being used within the doorway for trimmer studs. Cut them to length, using the cut-off portion for cripples above the doorway. The other cripples might be salvaged from other scrap.

•**One or two extra 8-foot 2 by 4 studs** depending upon where the doorway will stand in relation to the regular full-length wall studs to frame the doorway on the outside of trimmer studs.

•**Two lengths of 2 by 4 and one length of ½-inch plywood** (4 inches wide) about 7 inches longer than the door's width to be nailed together as a header.

•**About 19 feet of 2 by 4 fireblocking** will probably be needed across the wall's center. If the wall will be covered with board paneling, add two extra rows of blocking (an extra 38 feet).

•**Ten 4 by 8-foot sheets of paneling material** or a comparable amount of board paneling.

•**One door with casing, trim, and molding** (see page 27).

Wall Paneling—Choosing, Installing

"Shingle the living room wall?" "Sure—why not? Shingles will give it style, eye interest, texture. Why not try it?"

Why not, indeed. Although not a traditional interior wall covering, shingles are decorative and fit within the range of a carpenter's duty. In fact, carpenters apply hundreds of wall-covering materials, from shingles to plywoods and plastic-coated hardboard panelings. Most materials fall into one of three main categories: solid board, plywood and hardboard products, or gypsum wallboard. Many are structurally rigid; some, like shingles and thin wall panelings, are not.

This section gives a brief rundown of the materials within the three main categories and discusses their installation. (For information on shingling interior walls, adapt the methods for exterior walls given on page 47, remembering that "exposure" is not critical.)

SOLID-BOARD PANELING

Just as you can find a variety of millings for outdoor siding, so can you discover many decorative variations of solid-board paneling. Although almost any siding wood could be applied to indoor walls, the millings most commonly used are limited to square-edged, shiplap, and tongue-and-groove boards.

Choice of one type generally depends upon the overall appearance the homeowner wants his wall to have. All variations of tongue-and-groove and shiplap boards can be applied either vertically or horizontally. Square-edged boards are usually applied vertically in a board-and-batten or board-on-board pattern (see exterior siding section).

No matter what the milling, boards may be rough or smooth. The grade of wood may be "clear" or "knotty." Rough or knotty boards give a room an informal appearance; smooth or clear boards are more formal. For more information on lumber types, see page 24.

SHEET PANELING

For fast and easy application, most carpenters favor sheet-sized paneling materials, such as plywood and hardboard.

A wide range of plywood and hardboard paneling variations is available at any lumberyard. Paneling is characterized by differences in texture, surface-veneer, and finish. Some panels are meant to be painted, some resemble real wood, and others are veneered with real wood that is either prefinished or meant to be finished by the buyer.

Typical panel sizes are 4 feet wide by 8 to 16 feet long. The most common dimensions are 4 by 8 feet.

Plywood Products

Of plywood's many different uses, perhaps the key one is for wall paneling. Almost any of the many varieties of plywood can be used for paneling. Among its different textures and styles are resawn, factory-finished, and hardwood-veneered. For a complete selection of plywood types and installation techniques, see page 25.

Plywood products made especially for use as interior paneling include prefinished, unfinished, or vinyl-faced styles, convincingly imitating plank paneling, marble, and even tortoise shell. Various thicknesses range from 1/8 to 3/4-inch. Standard thickness is 3/8-inch.

Hardboard Products

Both prefinished and unfinished hardboard pieces make durable wall panelings. Even though prefinished panels are by far the most popular, the unfinished variety is often used where surface appearance is not a primary consideration.

Prefinished hardboard panelings usually have realistically-simulated grain and hardwood coloring. They are less expensive and more durable than real woods.

Panel surfaces can be obtained from the factory finished in plain colors, coated with hard plastic, or covered with vinyl.

GYPSUM WALLBOARD

By far the most popular wall and ceiling paneling material, gypsum wallboard is used as both a finish material and an underlay for other paneling materials. Two main kinds are available: a standard type that serves either as a complete, finished wall or as a backerboard, and a predecorated variety that makes an attractive finished wall.

Panels are 4 feet wide; their lengths run from 8 to 16 feet. You can choose from among several thicknesses—1/2-inch for final wall covering, 3/8-inch as a backerboard, and 5/8-inch on walls between garage and living structure.

Panels can be joined so that edge seams show (this is done with most decorative and backerboard types), or they may be taped and cemented to conceal joints (this is done with standard panels having tapered edges). For the inexperienced wallboard installer, taping and cementing joints well can be a considerable challenge (see illustrations on page 53 for techniques).

APPLYING PANELING

Methods of applying paneling differ with the various kinds, but the following information will be helpful when working with all types. Given first are a few miscellaneous tips and then some general working instructions. See the chart on the next page for details of installation.

- Handle materials carefully, most

TIPS FOR APPLYING PANELING

<table>
<tr><th rowspan="2"></th><th rowspan="2">BOARD PANELING</th><th colspan="2">SHEET PANELING</th></tr>
<tr><th>Plywood & Hardboard</th><th>Gypsum Wallboard</th></tr>
<tr><td>Cutting methods</td><td colspan="2">Use a saw. Cut paneling face up with a handsaw or table saw, face down with a circular or saber saw. Choose a fine tooth blade. Cutting edges of boards at a slight angle (about 5°) makes them easier to fit into place. This angle should slope downward toward the board's back side.</td><td>Use knife and straightedge. Score along cutting line with knife, extend the small portion over an edge, snap downward, then score paper in back of cut and snap upward. Use a saber saw or compass saw to make cut-outs for wall fixtures, or score cut-out on both sides and from corner to corner with knife, then punch it out with hammer.</td></tr>
<tr><td>Minimum backing necessary for nail fastening (see sketch, next page)</td><td>Place backing at sides, top, bottom, and provide enough intermediate backing to prevent boards from bending when heavy pressure is applied.</td><td colspan="2">Support around all edges and intermediately on 24-inch centers vertically or horizontally. For 8-foot-high stud wall, fasten one row of fireblocking between studs 48 inches up from floor to serve as nailing base.</td></tr>
<tr><td>Fastening

All nails should penetrate at least 1 inch into nail backing.</td><td>Tongue-and-groove narrower than 6 inches: blind-nail through tongue at every backing support. All others: face-nail twice per backing for boards 6 inches or narrower, three times for wider boards. Countersink finishing nails and fill nail holes if desired.</td><td>To flat surfaces, studs, or nailing strips: use a bead of adhesive along each backing and nail around borders to be covered with moldings (apply bead of adhesive every 16 inches on flat surfaces). Follow label directions of adhesive. Some panels have special fastening systems; follow manufacturer's directions.</td><td>Regular and backerboard: fasten with wallboard nails from center of each panel outward, every 7 inches along each bearing. Nail 3/8-inch from edges. Slightly dimple panel face with last hammer stroke on each nail head. Conceal joints and nail-dents of regular gypsumboard using methods shown in sketches. Predecorated gypsumboard: use adhesive, then nail around perimeters (as discussed for plywood and hardboard) with color-matched nails. Don't dimple surface nailing where molding won't cover.</td></tr>
<tr><td>Application steps

First panel should be plumb (vertical) or level (horizontal) before applying the next.</td><td>Horizontal: beginning at bottom, trim ends of boards to fit. Working from bottom to top, you may need to fit each board if adjoining walls are uneven. If boards are shorter than wall's length, join them only on nail backing. Vertical: beginning at one corner, scribe board to fit adjoining wall if necessary. The job of fitting consecutive boards will be standard (8 feet) unless floor or ceiling is uneven. Check measurements frequently during application.</td><td>Apply decorative panels vertically, regular panels horizontally beginning at one corner. If necessary, scribe to fit uneven adjoining surfaces. End-join and edge-join panels only on center of nailing members.</td><td>Regular and backerboard: apply horizontally, working from a top corner down and across. Scribe to fit if necessary. Lower boards should be snug but not wedged into place. Apply corner moldings, and conceal joints and nail-dents of regular panels after all panels are installed. Predecorated gypsumboard: fasten vertically, beginning at one corner and working across the wall. Preliminary application of backerboard is usually recommended.</td></tr>
</table>

can be easily damaged.

•Board paneling should be delivered about a week early and stored in the room to be paneled. This will give the wood a chance to attain the same moisture content as the room, minimizing shrinkage once the paneling is installed.

•Carefully figure the placement of paneling before starting work. Do this so that the last board or piece of sheet paneling you fit into place won't be too narrow and so that panel edges match up to doors and windows whenever possible, limiting the number of necessary cuts.

•Both adhesive and nails are used for installing panels (see chart). Nails should penetrate into studs at least 1 inch. When nailing finish surfaces, use colored paneling nails or set the nails and fill holes.

Special paneling adhesive is available in most building supply stores. Follow manufacturer's directions carefully.

•Different wood moldings available are discussed on page 27. Most carpenters suggest using moldings only where needed to conceal unsightly cuts or nail marks around the perimeter of a room.

TO TAPE GYPSUM WALLBOARD, first apply smooth bedding coat of compound over all joints and nail dents with 6-inch drywall knife. Then, before dry, smooth tape over joints; apply thin layer of compound over tape. When it dries, apply second coat, feathering edges. When that dries, apply last coat with 10-inch knife, feathering edges. Coat nail dimples during each stage of joint concealment.

HOW TO FUR A WALL

Whether you're planning to panel directly over open studs, over a previously covered wall, or over masonry, you need a sturdy base for fastening paneling. Wall studs serve as an ideal fastening base. The need for extra horizontal blocking between studs depends on the type of paneling.

Apply paneling to smooth, flat, previously covered walls by nailing through the covering into studs or fix panels directly to the covering with adhesive.

For masonry walls, walls that are not flat, or in places where you simply need more fastening support, attach 1 by 4 furring strips as shown. You can nail the strips to wooden walls. Be sure nails always penetrate into studs at least 1 inch. If you need help in locating the studs beneath

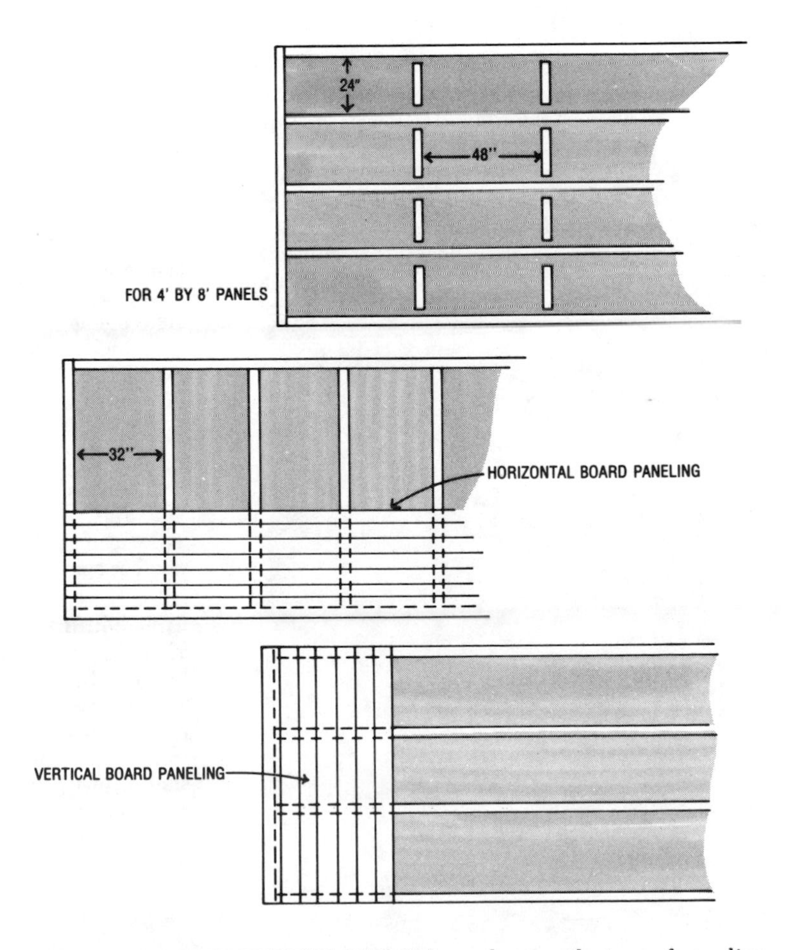

ARRANGEMENT OF FURRING STRIPS depends upon the type of paneling to be applied to the wall. Be sure to apply strips firmly to wall.

a wall covering, see below.

Fasten strips to masonry walls using concrete nails or expansion bolts (page 32).

Follow the grid patterns shown for different paneling materials.

HOW TO SCRIBE PANELING

Probably the first piece of paneling that you fit into the corner spot of a wall will not fit exactly the contours of the adjoining wall or floor. Nor is it likely to be level or plumb. To overcome this you must duplicate the irregularities of the adjoining surface on the paneling's edge. An inexpensive scribing tool is made especially for this purpose. Costing less than $1.00, it is shaped like a short compass. One leg has a pointed shaft; the other, a pencil.

Prop the panel into place about an inch from the uneven adjoining surface, using shingles if necessary to shim the panel into level or plumb. Holding the scribe's points parallel to each other, draw the scribe along the surface so that the pencil leg duplicates the uneveness onto the paneling.

Typical tools used to cut paneling along scribed lines include the coping saw, block plane, chisel, or knife. The size of cut needed will determine the tool you choose. Don't use any type of plane on gypsum wallboard.

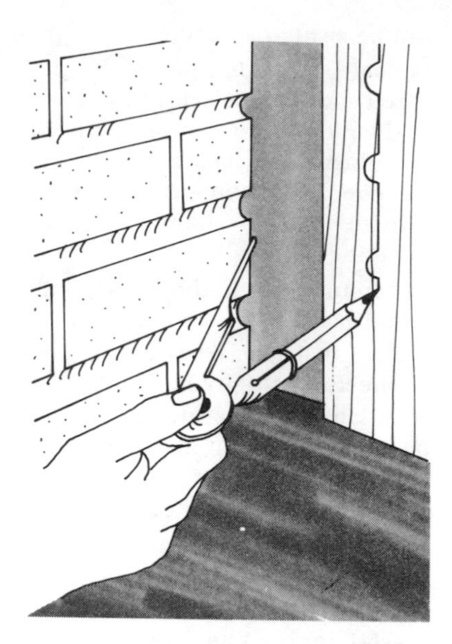

HOW TO FIND A WALL STUD

One problem you may encounter when working with walls that are covered with paneling or siding is finding the location of studs beneath. If you run into this difficulty, first gain a basic understanding of wall structure, as shown on page 41.

One thing you'll notice is that studs should be spaced regularly, except for extra studs around windows, doors, and at corners. Although studs are usually spaced on 16-inch centers across the wall's length, sometimes 20 or 24-inch spacings are used. With this in mind, finding one stud 16, 20, or 24 inches from a corner will probably enable you to locate the rest by measuring.

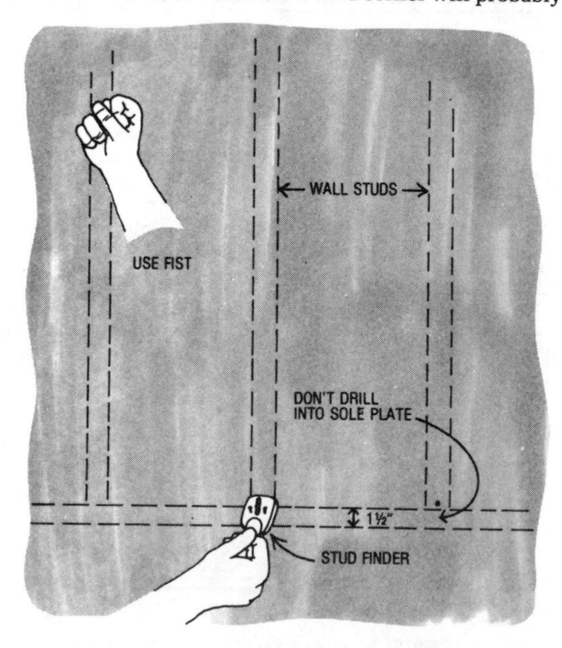

You can often see the nails that hold wall-paneling materials to studs. If you don't see any nails, check for those that are holding paneling or siding to the other side of the wall, measure their location from a corner or doorway, and transfer the measurements to the right side of the wall.

Nails in gypsum wallboard are probably covered with special tape and compound. If you can't see nails but know they are there somewhere, you can buy an inexpensive commercial stud finder to locate them. This handy little tool has a magnetized needle that, when passed over a nail, will signal its presence by fluctuating. If the wall paneling was applied with an adhesive, you'll have to try another method.

Knocking firmly on the wall with a knuckle will sometimes do the trick. Rap sharply in the area where you think a stud is located. A solid sound means a stud is beneath; a hollow sound means "Knock again in another spot."

If all else fails, make exploratory holes into the wall using a drill and small bit or a small nail. Either drill a hair's breadth above the base molding and fill the holes later or take up the molding and drill where it will cover. The bit or nail will penetrate the paneling material and then either reach solid wood or hollow space. If it reaches solid wood, this means that you have either found a stud or just penetrated into the sole plate (which may rise 1½ inches above the floor). Make another hole about 3 inches to the side to be sure you're not just drilling into the sole plate. If you hit the sole plate, make a hole slightly higher. After you are sure you've found a stud, try locating the rest by measuring and knocking.

Building into a Wall

By building into a standard wall, you can gain an extra storage compartment without sacrificing any floor area. Such space is especially helpful for storing small items that would tend to be swallowed up by an ordinary shelf.

Because the shelves of a built-in compartment are necessarily shallow, you can usually stack items only about one deep. Illustrated are some examples of areas in the house well suited to the installation of inner-wall storage compartments.

Be sure that the spot you choose for one of these units is away from electrical switches and outlets, plumbing, and heating vents so that you won't open up an area that contains wiring or other fixtures.

Because wall studs are usually on 16-inch centers, you can usually find them by measuring out from a corner of the room. If that doesn't work, however, refer to page 54.

For methods used to remove wall covering materials, see page 62.

If you're putting in a narrow cupboard, such as a medicine chest, you won't need to cut away any sections from the wall studs. Each of the compartments shown here is wider than the 14½-inch distance between two adjoining studs, though, so installing the compartment involves not only cutting into the interior wall but also cutting away part of at least one stud.

Mark with a square any cuts you may need to make within an opening. Then, working carefully, use a handsaw to cut studs. Be sure to allow for a 2 by 4 sill and a header; most building codes require a 4 by 4 header over openings between 16 and 48 inches wide (for wider openings, see chart on page 62).

If you're planning to install a wide cupboard that calls for the removal of parts of more than two studs, it's advisable to put props under the ceiling before you cut to take the load off the studs. Place a length of 2 by 4 flat against the ceiling several feet out from the wall,

A HANDY BOOKSHELF FOR KNICK-KNACKS, when built into the wall, will provide plenty of storage space yet require very little floor space. Another idea you might try, is a magazine and book rack. You can locate one of these over the bed as shown, or in almost any other room.

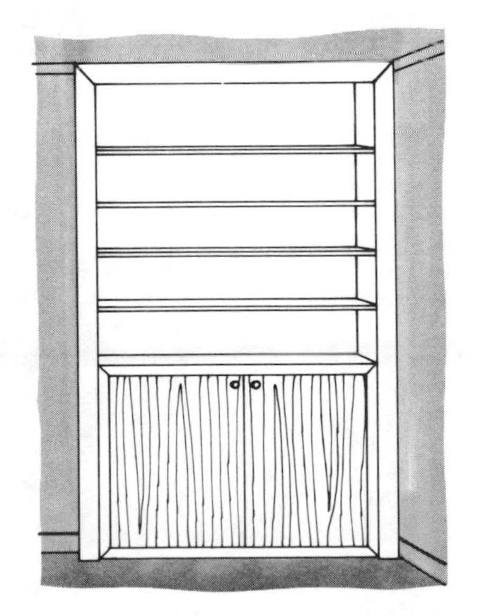

TWO OTHER SPACE-SAVING IDEAS are illustrated here. The first one is a narrow cupboard that you can build either flush to the wall or extended out with a slightly wider jamb. The second is a handy telephone desk. Its surface is the back side of a plywood door that swings down.

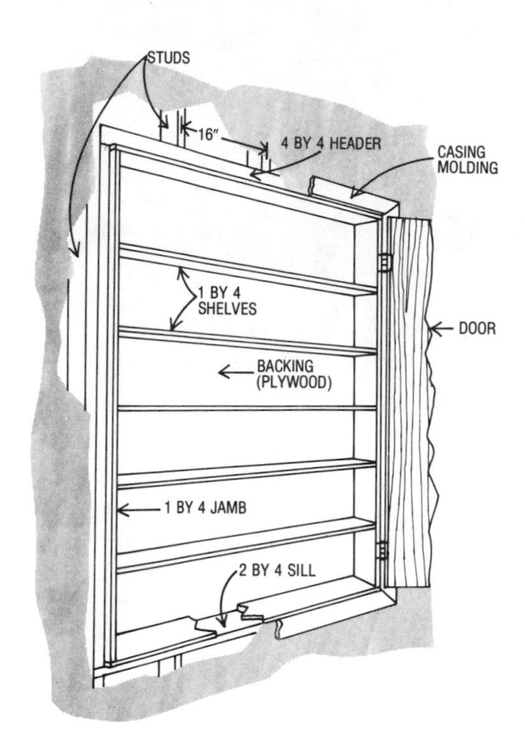

propping it up with three or four 2 by 4's while you trim the studs and put in a header (or see the support shown on page 66).

Toenail header's ends to full-length studs and toenail through cut-off studs into header's top. Nail cripples to studs to support the header's ends. Then toenail sill's ends to full-length studs and nail through sill into the cut-off studs. For further explanation of these steps, see procedures for rough framing windows on page 68.

Construct the compartment framework as shown in the sketch, modifying it to fit your needs. The molding should be wide enough to cover the raw edges of the wall material. Nailed to the back edges of the 1 by 4's, a back that's the same thickness as the wall covering will bring a 1 by 4 jamb out flush with the wall covering.

If you want a deeper cupboard, use a wider jamb and let it extend a couple of inches out from the wall.

Put the cupboard between the studs; nail through the jamb to fasten the cupboard to the cripples, header, and the sill. Then add a casing to conceal the gap between the jamb and the wall. If you want doors on the cupboard, hinge them to the jamb.

OR DO YOU NEED A PASS-THROUGH?

You can use these same steps described above for building a pass-through into a wall—just take the wall covering off both sides of the wall, use side and top jambs wide enough to extend flush with both walls, and attach a wide, counter-sized unit instead of a bottom jamb.

POST AND BEAM CONSTRUCTION

If none of the framing methods described so far resemble the framing in your house, it may be framed using "post and beam construction." An increasingly popular method of building houses, particularly in the West, post and beam framing is a simple method, keeping building costs considerably lower than standard framing. Not only is post and beam framing inexpensive but also this popular method lends variety to architectural styles.

Post and beam-framed houses have heavy members at long intervals instead of light members at short intervals, as in standard framing. Because of this, large scale remodeling is often very difficult to do. Each post and beam-framed house usually requires all of its posts and beams; these elements hold the house together. You should never remove a post or beam without professional assistance.

You can usually tell quite easily if a house is framed with posts and beams. Posts are often exposed between sections of wall paneling or exterior siding. Some of the rooms may also have ceilings of exposed beams. Plank roofing and flooring materials, in conjunction with their structural framing, often eliminate the need for rafters and sub-ceilings.

Although standard wall paneling may fill the gaps between posts, many post and beam houses have special prefabricated panels (sometimes bonded with insulation) that serve as walls when fastened between posts. It's advisable not to attempt changing such panels. Instead, remove them completely and reframe the opening to suit your purpose as shown in the sketch.

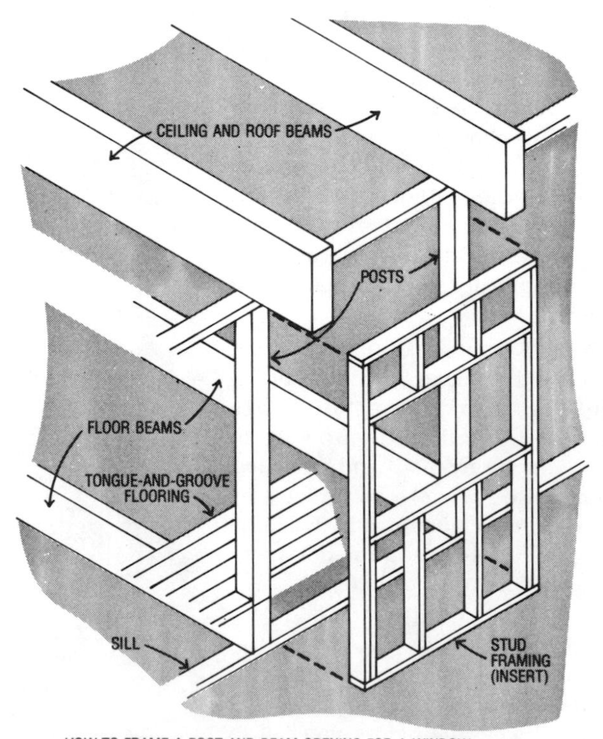

HOW TO FRAME A POST AND BEAM OPENING FOR A WINDOW

Bookshelves You Can Try

Several great authors and poets have found their final resting place in London's Westminister Abbey. The spirit of their works, though, may find a set of shelves in your living room a more fitting place to lay in respose.

Bookshelves—needed by almost everyone—are fortunately quite easy to build.

Sometimes shelving is inset within a wall, as discussed on the previous two pages; other times it's attached to a wall (fasteners, page 31) or built freestanding. Unless made from exceptionally sturdy materials and design, a freestanding bookcase usually requires a back.

The ideas on this page are given to help you in designing bookshelves for your specific needs. These suggestions are divided into three design types: (1) shelving locked together with a back and support intermediately and at each end of shelves, (2) component or "modular" shelving, in which case the size of the bookcase is determined by the number of units you make and assemble, and (3) adjustable shelves that are supported only at each end. This shelving unit is attached to the wall.

Remember that these bookshelves are only design ideas from which you can create any number of variations. For instance, in the case of number (3), the concept is that supporting elements hold the shelving boards at each end. You may decide to make these supporting elements from tall lengths of 2 by 8, from stacks of concrete blocks, or from 4 by 4's and dowels, instead of the

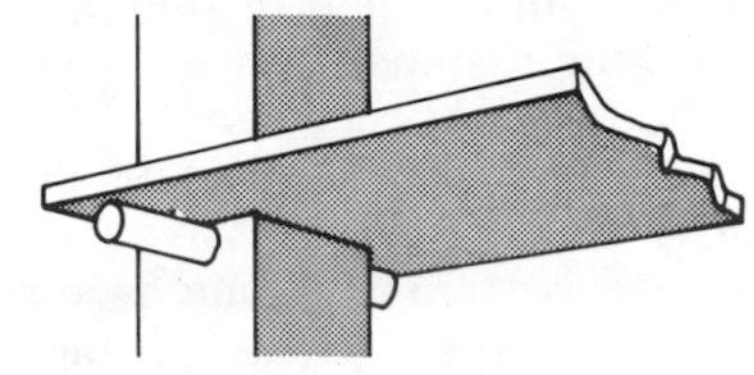

2 by 3 and dowel assembly shown.

Whichever approach you choose, let your imagination help you create a unique bookcase that fits your particular needs.

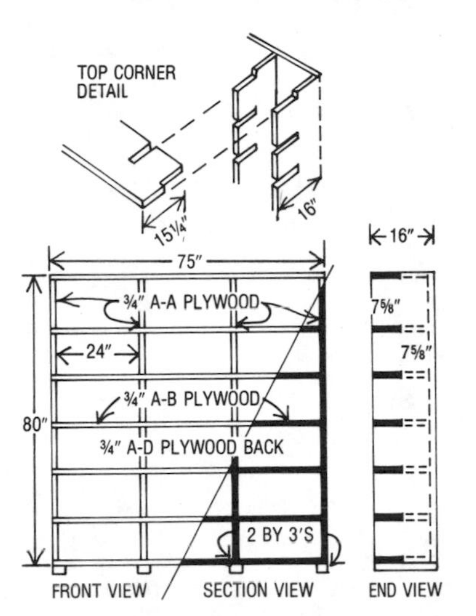

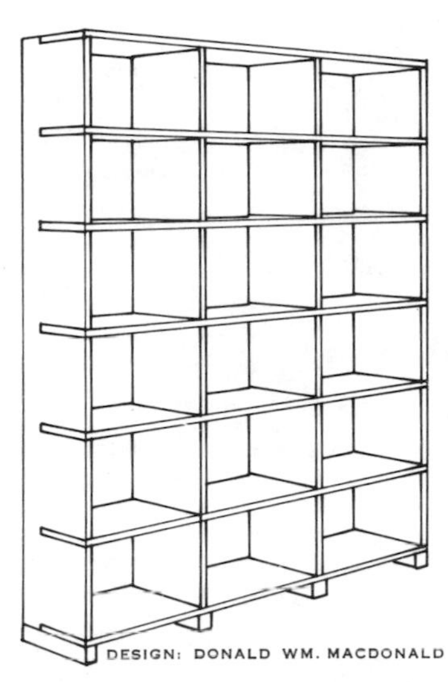

PLYWOOD SECTIONS, slipped together, form this bookcase. Once you've marked and cut one horizontal and one vertical panel, use them as templates to mark the other pieces. Base is 2 by 3's on edge.

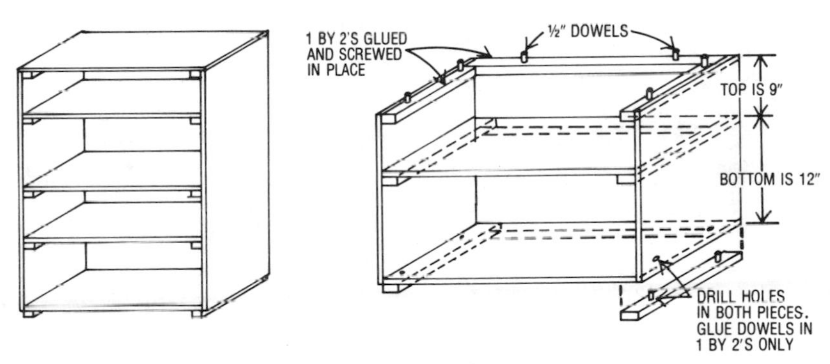

MODULAR SHELVING BOXES fit together with dowels. Glue and screw 1 by 2 shelf supports to the ⅝-inch A-D plywood sides and back. Inset shelves far enough to leave room for plywood edge treatment (page 26).

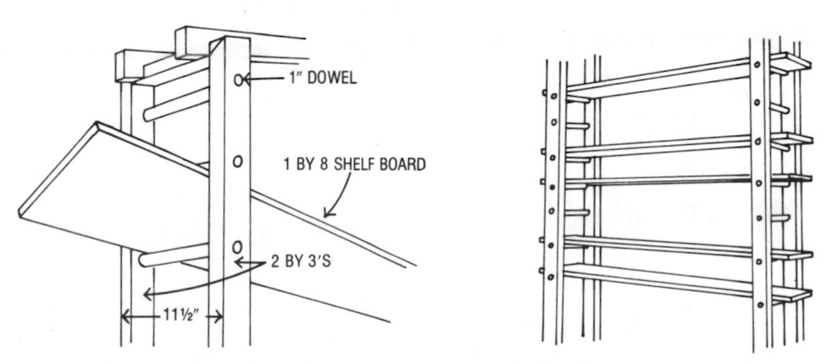

DOWEL RUNGS (1-inch) and fir 2 by 3's make the ladder-type supporting framework for these shelves. The shelving boards are 1 by 8's. Lock the two ends together with notched 2 by 3's.

Ceilings

Imagine crowds of people wandering through your home gazing at its ceilings as if they were in the Sistine Chapel. You say you'd rather not imagine that? Don't feel alone. The ceilings of most homes are not their most outstanding feature.

Even so, a handsome ceiling can brighten up a room with light, color, texture, and a clean, finished appearance. So if the ceilings in your home are in ill repair, or rooms are capped by nothing but open joists, they deserve attention. The time you'll spend giving new life to your ceiling will hardly compare to Michelangelo's years of frescoing.

The work involved in putting up a ceiling depends largely upon the ceiling material you choose. Suspended-panel ceilings and ceilings of mineral and wood-fiber tiles are easily installed. Gypsum wallboard and wood paneling are more awkward and complex to work with.

First in this chapter you'll find basic information on typical ceiling framing to help you gain a full understanding of the structure you'll be working with. Next, you will see an illustrated guide for installing the two easiest types of ceilings: suspended panels and ceiling tiles. Last, a few tips are offered to make the work easier for those who choose to apply gypsum wallboard or other products to ceilings.

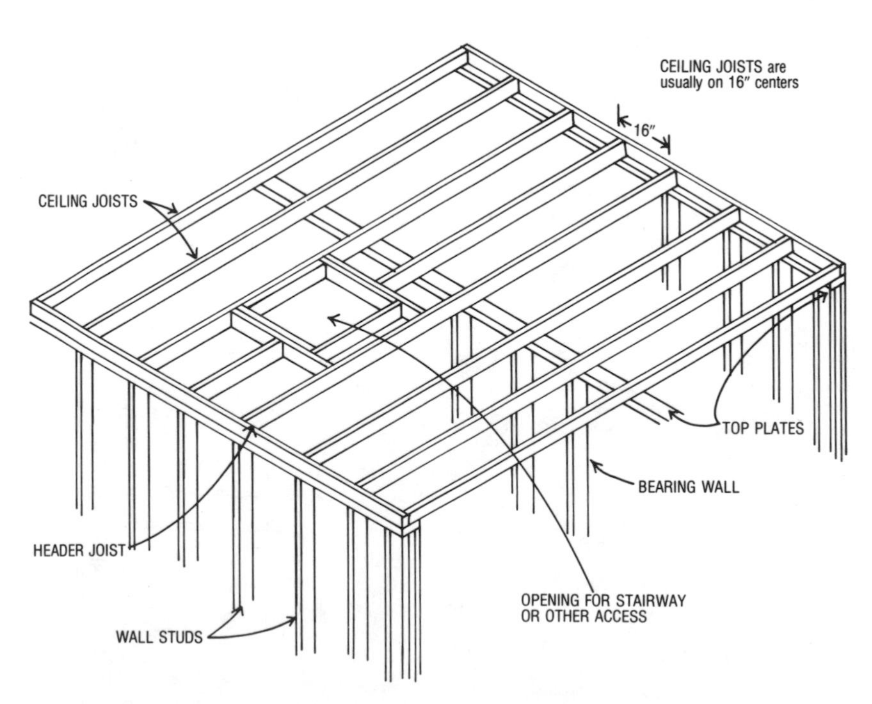

CEILING FRAMING of most houses are variations on the basic pattern shown here. The ceiling frame of one room may be the floor frame of the room above.

THE CEILING'S FRAME

The way a ceiling is framed depends on what's above it. Framing for a ceiling on the top story of a house is not the same as the framing for a basement ceiling. Whereas the top story has special framing, the ceilings of stories below are carried on the underside of floor joists. The ceiling framing of a first story is the floor framing of the second story).

As explained in the chapter on floors, the primary framing members are called "joists." Their sizes depend on the length they must span, the distance between them, and the weight they carry in their other role as floor joists. Attic floor joists, if not carrying a "live load," are not as large or strong as joists on other floors. Typical spacing of all joists is 16 inches on center.

All joists must have a strong bearing capacity at both ends. Usually, but not always, they run across the narrow dimension of a house or structure. Sometimes one series of joists will run in one direction and a second series will connect to the outer member of the first series at right angles.

A single joist is seldom one length of lumber; rather, it is usually made up from two or more lengths of lumber lapped or butted and nailed together. The midpoint where two joist lengths join must be supported either by a beam or by a bearing wall.

HOW ABOUT A SUSPENDED CEILING?

A suspended panel ceiling is ideal for a room where you want to lower the ceiling or hide unsightly joists, rafters, or fixtures. Easy to install, one of these ceilings usually consists of a metal grid suspended from above with wire or spring-type hangers that hold acoustic or decorative fiberboard panels. Transparent, translucent, or egg-crate grilles are made to fit the gridwork in place of the fiberboard panels to admit light from skylights or lighting fixtures. All of the components are replaceable, and the panels can be raised for access to the area above at any time.

You can decide on color, texture, and overall appearance of both panels and supporting grids. Choices include acoustic or decorative paneling in several textures, matching or contrasting supporting grids, and gridwork that is either obviously exposed or somewhat hidden between panels.

Some insulating-acoustical ceiling systems have panels as large as 4 by 16 feet, but the most commonly-sized panels for suspended ceilings are 2 by 4 feet.

Here is the easiest way to figure the number of panels you need. Using graph paper (or ruling a plain sheet of paper into equal squares), measure wall lengths at the proposed ceiling height and draw the ceiling area on the squared paper to scale, using one square per foot of ceiling size. Block-in the panel size you will be using. Then count the blocked areas and parts of areas to get the number of panels you will need.

For a professional-looking job, make the opposite borders of the room equal. To determine the nonstandard width of panels needed for perimeter rows, measure the extra space from the last full row of panels to one wall and divide by two. This final figure will be the dimension of border tiles against that wall and the wall opposite it. Repeat this procedure for the other room dimension. Use the same procedure for figuring ceiling tiles.

Shown are the steps for installing a suspended-panel ceiling.

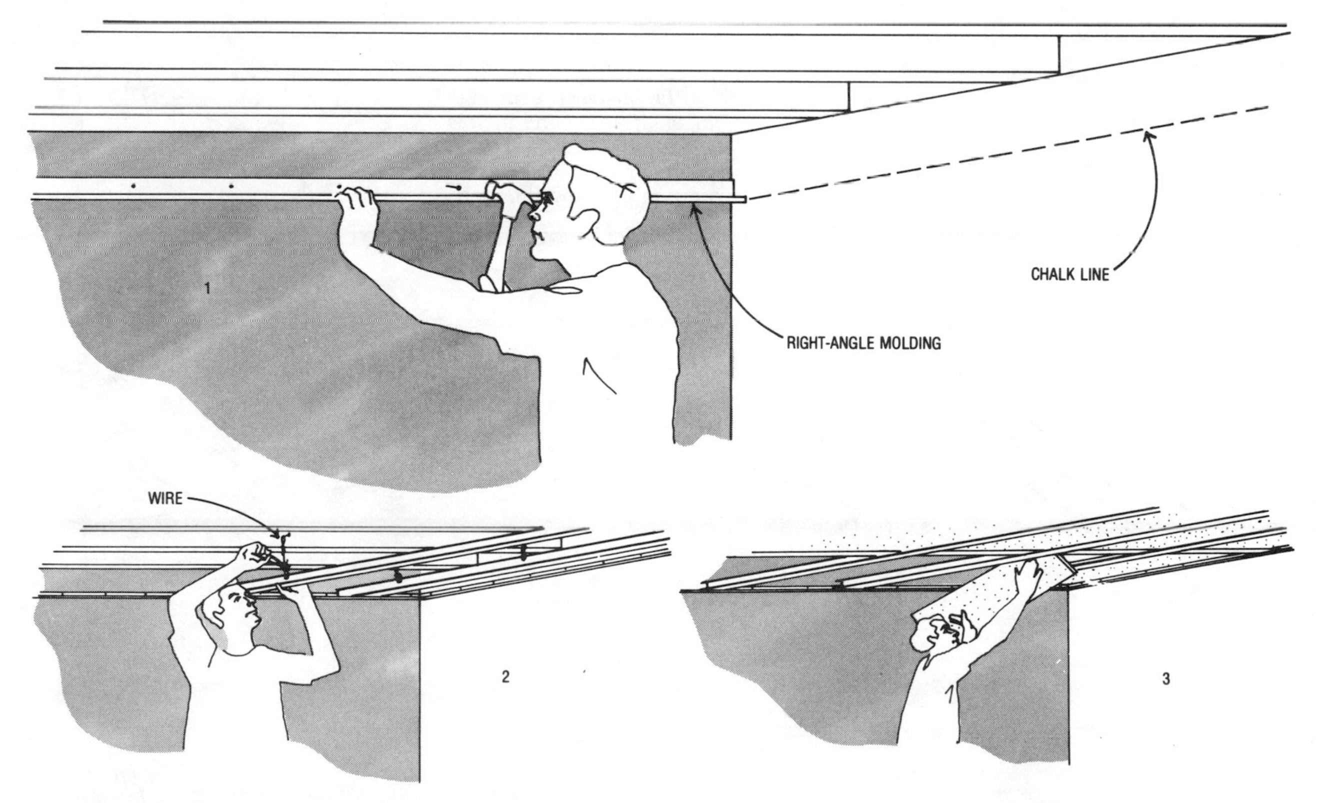

INSTALL A SUSPENDED CEILING following these steps. (1) Figure the ceiling height at least 3 inches below plumbing, 5 inches below lights. Snap chalkline around room at that level and install L-shaped angle molding with its base on chalkline. (2) Install main runners perpendicular to joists. Cut them to length with tinsnips. Setting them on molding at each end, support them every 4 feet with No. 12 wire attached to joists above. (3) Lock 4-foot cross tees to the main runners by bending tabs in runner slots, and push panels into place. Cut panels with knife or fine saw.

1

CEILING TILES need a grid. Nail the first furring strip flush against wall at a right angle to ceiling joists.*

2

MEASURE CENTER of second strip one tile width from wall. Then space other strips, center to center, by the width of one tile, checking about every 4 feet with a tape measure for accuracy.

3

LEVEL strips by driving shingles between the strips and joists. Strips should also be level with each other.

4

STAPLE border tile through flanges after cutting to size with a fine-toothed saw or a knife. Face-nail to furring where molding will cover.

5

ONE EDGE of tiles should be centered on the nailing strips. Do border tiles at wall first, then work outward across the room.

ON FLAT CEILING, install tiles by daubing special adhesive on corners and center of each tile's back.

INSTALLING CEILING TILES

Square and rectangular ceiling tiles are available in several decorator and acoustical styles. These tiles, usually 1-foot square, can be applied either directly to old ceilings or to 1 by 3 furring strips that cross joists or ceilings. Either adhesive or staples are used for the bond.

Estimate the number of tiles needed and the dimensions of the odd-sized perimeter tiles by the same methods used for suspended-ceiling paneling (given on page 59)—merely substitute "tile" wherever "panel" is mentioned.

If you plan to apply nailing strips first but joists are covered with an old ceiling material, locate them using the same methods used for locating wall studs given on page 54. They are usually on 16-inch centers. Mark their locations on the old ceiling with a chalk line so that you can find them rapidly when nailing strips in place.

GYPSUM WALLBOARD FOR CEILINGS

Gypsum wallboard is not one of the more easily-installed ceiling materials. The panels are large, awkward, and quite heavy, making them hard to hold in place while nailing. In addition, concealing the joints between panels to yield a smooth surface typical of gypsum wallboard is an exacting job.

Methods of ceiling application are basically the same as those for walls given on page 52. You need a flat, true nailing surface. If the old ceiling or bottoms of joists above are uneven, apply 1 by 3 nailing strips perpendicular to joists on 24-inch centers. Apply panels perpendicular to joists over these strips.

Because it is necessary to hold the large panels up while nailing, this is a two man job. Set up a couple of sturdy sawhorses, laying a few large planks across them to serve as a short scaffold to stand on.

The procedure is this: both men—hammers and sheetrock nails in hand—hold their respective ends of a panel in place with their heads. Begin nailing at the center of the panel and work outward. Space the first few nails where they will take the weight off your heads. Nail spacings are governed by codes.

If the room is being paneled in gypsum wallboard, the ceiling should always be applied before the walls. The corner joint between ceil-

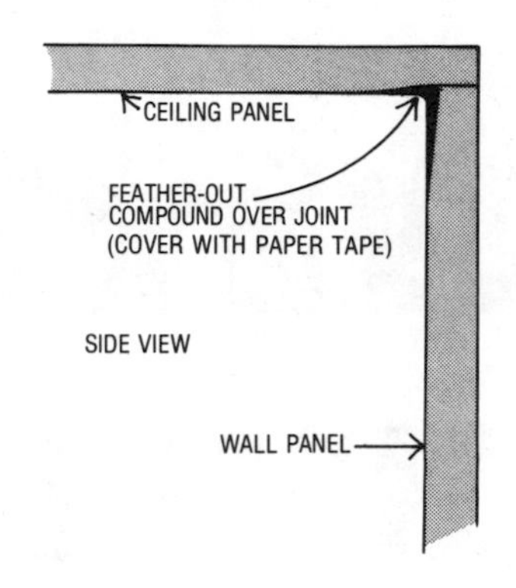

ing and wall gypsum boards is formed as shown.

OTHER CEILING COVERINGS

A host of wall covering materials may be used to cover ceilings (after all, a ceiling is much like an upside-down wall). As with gypsum wallboard, the chief difference in application lies in the handling of these different materials.

Among those wall-covering materials used for ceilings, you can choose from solid-board panelings, plywood, and even some types of sidings. To apply these materials, simply adapt the directions given for their application to walls in the chapter having that title.

IT TAKES TWO to install a gypsum wallboard ceiling. Prop each panel into place with head and nail from center outward.

Windows and Doors

Doors and windows are like the eyes, ears, nose, and throat of a house. Windows let in sights, sounds, and smells; doors let in the staff of life—people. As a rule, homeowners enjoy having many windows and doors, enabling them to keep closely in touch with the world outside.

Many home carpenters have a project in mind requiring the addition of a new door or window. For instance, when building a deck, indoor living is also enhanced by adding a sliding glass door in order to bring the outdoors in. Or a home owner adding a closet or simple wall may wish to include a standard door. Of course, an extra window or two is almost always desirable for adding light or expanding a view. These are the types of projects that this chapter examines.

Opening Up a Wall

Ideally, you should frame for a window, sliding glass door, or standard door during the construction of a wall (page 48). But perhaps you weren't around when the wall was built or since that time have discovered a need for more light or easier access to the patio or deck. In either case, you'll need to open up the wall and reframe it to receive the window, sliding glass door, or standard door that you desire.

Once the wall is opened, cleared of unwanted studs, and reframed, the procedures for hanging a door, window, or sliding glass door are similar to installation in a newly-framed wall.

Because opening up the wall can be an exacting job, it is a good idea to read through the procedures given for your particular project before deciding to do all of the work yourself.

Given here are some general tips, methods for locating the opening and removing the wall's skin, and a little information on extracting studs, adding headers, and so forth. Look to particular sections within this chapter for specific measurements and directions.

Before You Start

Your first step should be to take a close look at the wall concerned:

Will the job involve this wall only or will it really lead you to remodeling or redecorating the whole room? You might as well face all the facts now, avoiding bills that run way over your original cost estimate.

Is the wall a bearing wall? Any exterior wall or interior wall at right angles to the ceiling joists should be considered a bearing wall.

In some cases, you should get professional help in determining header size, especially if there is a floor load above it. The table at right indicates header sizes for most openings.

Be sure to order your header at least 2 feet longer than the net length of the opening to allow for trimming after you determine the exact stud spacing. Plan to have a helper on hand when it's time to install the new header.

Will opening the wall require any rewiring or relocating of plumbing pipes and fixtures?

If you need an electrician, have him there on the day you knock out the first side of the wall.

To determine header size . . .

Opening Width	Header Size
Up to 4′-0″	4x4 or 2-2x4's on edge
4′-0″ to 6′-0″	4x6 or 2-2x6's on edge
6′-0″ to 8′-0″	4x8 or 2-2x8's on edge
8′-0″ to 10′-0″	4x10 or 2-2x10's on edge
10′-0″ to 12′-0″	4x14 or 6x12 on edge

Specify Douglas fir or Southern pine.

Sizes meet most local building codes.

For wider openings or other woods, consult an engineer.

Relocating plumbing can be complicated and expensive. Check on plumbing pipes or venting stacks by looking both under the house and in the attic. Even though the fixture itself is on another wall, you may find a vent in a wall. Before knocking out any wall with plumbing, have a plumber estimate the cost of relocation.

You may have to take out a building permit, depending on where you live and the extent of your job. Call your city's building inspector to find out whether a permit is required for your project.

Locating the Opening

If you have any choice in the matter, you may be able to move the opening to one side or another to make your job easier or to avoid wiring or plumbing. Check through these other tips:

- In a house with ceilings of conventional height (8 feet to 8 feet 3¼ inches), you will probably prefer having the top line of doors and windows 6 feet 8½ inches up from the floor. In height, this line will match other windows and doors in the room, and you can install a header of almost any size between the ceiling and door and window tops.
- Don't forget to look at the wall from both sides. An opening centered in a wall may look all right from one side but may not be the most effective solution when you are looking at the other side.

Removing the Wall's Skin

It's easy to knock down a wall, but there's an art to knocking out an opening that will be easy to patch or redecorate. If you take it slow and easy, your clean-up job won't be nearly so time consuming.

Before beginning work, cover the inside floor with a large tarpaulin (frequently empty it outside as you work).

Because the inside wall is usually the easiest to patch and refinish, cut into it first. In addition to the dimensions of the opening you want, you must expose enough of the framing to allow placement of a new header and trimmer studs (refer to diagrams on the following pages). In some cases—when patching looks like a bigger job than total replacement—remove the entire wall covering.

Inside wall. You'll probably be able to saw gypsum board or any wood-product paneling to an accurate line without damaging the surrounding wall surface. Either cut with a power saw or bore a pilot hole just inside your outline with an old auger bit and use a keyhole saw until you have an opening large enough to insert a crosscut saw. Look for the taped joints of gypsum wallboard or joints of paneling if you plan to remove a considerable amount: you may be able to remove one or more 4 by 8-foot sections intact. Use a wrecking bar or the claw end of your hammer, prying gently.

If your interior wall is plaster, first use a broad-bladed cold chisel and hammer to chop out a horizontal ribbon of plaster between two laths as close as possible to the horizontal outline you've drawn.

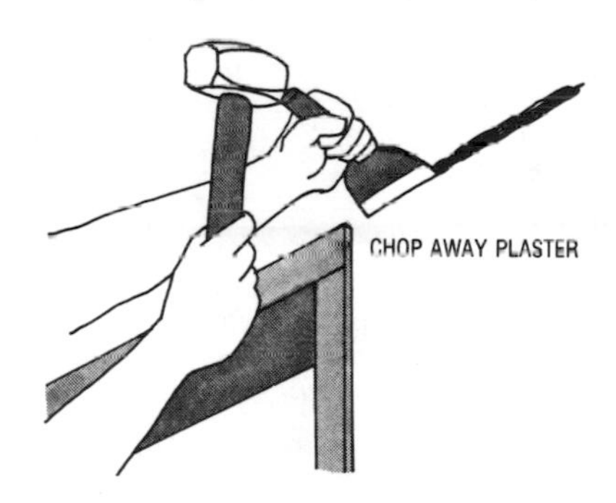

Next, use the cold chisel and hammer to chop out several inches of plaster just inside the vertical lines you've drawn. This exposes the laths (or button board), and by sawing through them as closely as possible to a stud, there is less chance of cracking or loosening the plaster. Again, you can use a power saw with a masonry blade for the entire cutting operation.

When the wall is opened up, you can confront hidden problems. It may be necessary to remove or re-route wiring, plumbing, bracing, and such. After carefully placing the header and new studs from the inside (see related sections that follow), you can precisely locate the exterior opening, minimizing patch-up work. (Drive long nails through from inside to mark final opening's four corners.)

If your removal job includes an existing door or window frame, you can often saw through the nails holding the frame in place (using special blades), then remove it in one piece after you've lifted out the actual window or door.

Outside wall. The most difficult common exterior wall material to cut and patch is stucco; shingles and wood siding are the easiest. All you need for removing siding or shingles are a hammer and saw (a portable power saw is very handy). Cut carefully—patching is not an easy job.

Old stucco will probably be keyed to chicken wire nailed onto the sheathing. To remove, mark your opening outline, then divide it into a grid of roughly 2-foot-square sections. Smash the stucco along these outlines with a hammer and cold chisel, exposing the chicken wire. Then use tin snips to cut the wire along these outlines and pull whole sections right off the sheathing. Use heavy leather gloves, for chicken wire can cut you. Better still, use a power saw with masonry blade to cut stucco and wire together—but wear eye protection.

Patching Up the Wall

When using wall-covering materials that can be cut to a straight line, patch-up work is relatively simple: just cut replacement pieces to circle the opening, applying them as you would a new wall. Caulk around edges (see page 34).

To repair stucco, mix a batch of one part Portland cement to three parts coarse sand, adding enough water to make a stiff paste. Apply in three coats (allowing 48 hours between each), gradually working out to the wall's level. Add ¼ part lime to the final mixture.

How to Install a Door

Framing and hanging a door once was an exacting job for the carpenter, but now that factory-built prehung door frames are on the market, the job can be quite painless.

Stud framing for a door should be like that shown in the sketch. If you're not starting from scratch, you'll have to cut through the wall covering material as described on page 63 (specific measuring directions follow), then cut away the studs within the proposed opening, changing it to the proper framing.

No matter whether you're cutting the door into an old wall or starting from scratch, you should have the new prehung door frame on hand when it comes time for measuring.

What Is a Prehung Door Frame?

The easy-to-install prehung door frame is like a box with a door hinged to it. This box is easily slipped into a stud-framed hole and then secured.

Each prehung frame is made up of two side jambs and a head jamb, dadoed together at the top. A molding, called a door stop, runs around the inside of the jambs to keep the door from closing too far.

Exterior doors need a sill and threshold at the base of the two side jambs. Usually milled from one piece of lumber, the sill slopes away from the door's base to keep water out, and the threshold closes the opening between the bottom edge of the door and the floor.

A prehung frame usually comes

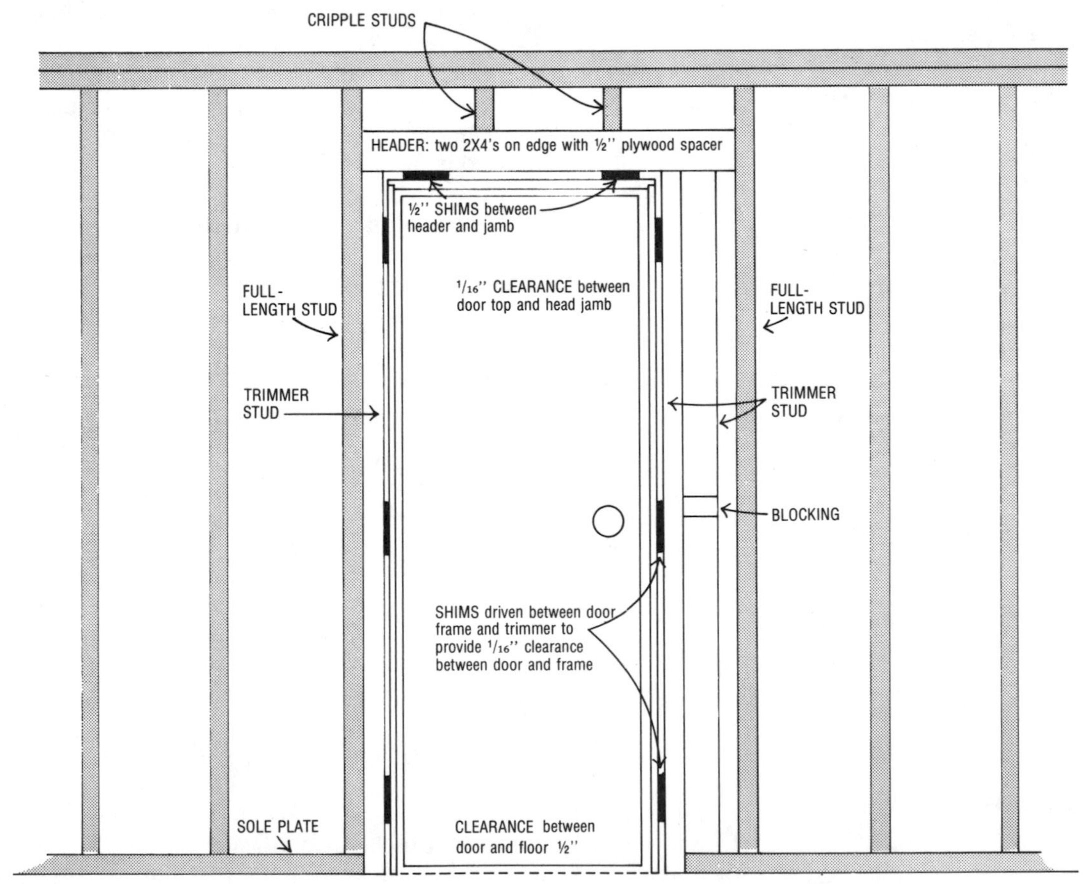

FRAMING A DOOR in a previously framed wall is shown here. Shaded studs are established; leave them and add new framing.

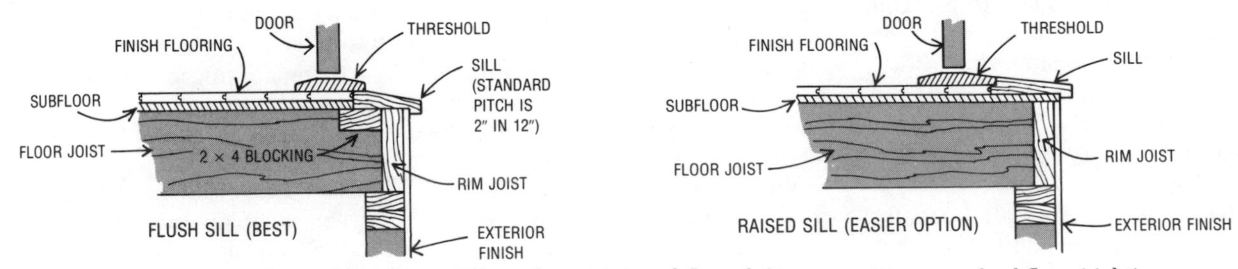

SILL AND THRESHOLD are for exterior doors. Sill can be cut into subfloor (left) or can sit on top of subfloor (right).

from the dealer with the door already attached on hinges and the door stop molding either tacked to the frame or bundled separately. You pull out the hinge pins to remove the door while securing the frame in place. If the door stop is tacked to the frame, also remove it before beginning to install the door frame.

The door may or may not be equipped with the latch assembly and casing (molding that hides a joint between wall surface and door jamb). If it lacks these, buy them separately.

Removing Wall-Covering

Properly laying out the location and measurements of the door is the first step for cutting the new door into an old wall.

The location of one particular line is very important: the line designating where the top of the new door's header will meet the cut-off ends of old wall studs. This will be the top line for cutting through wall covering and will govern where to sever the old studs.

Figure this line by adding the header's width (3½ inches for two 2 by 4's on edge), plus ½-inch for shimming between frame's head jamb and the header, plus the prehung door frame's finished height.

Some prehung door frame's side jamb bottoms are trimmed on the job to allow ½-inch clearance between door bottom and floor. If you've purchased one of these frames, trim it before figuring the dimension (don't forget to consider the thickness of any flooring material not already laid).

Plan sides of doorway so that one trimmer stud can stand against an existing full-length stud. The other trimmer stud can be blocked out from a trimmer nailed to its nearest existing full-length stud to provide the proper openings (see sketch).

When cutting the wall covering off, remove it to those existing full-length studs on the outside of each trimmer stud's proposed location.

Cutting Away Unwanted Studs

After removing the wall covering, the next step in reframing an old wall for a door is to cut away the unwanted studs located in the proposed doorway.

First, cut the studs about 3 inches up from the sole plate (to avoid nails). Then cut carefully along that important top line established during wall-covering removal. To insure accuracy, it's best to mark each stud for those top cuts with a square. Stud ends must butt firmly and squarely against header's top.

A Door's Proper Stud Framing

The first rule to remember when framing for a door is that its framing must be square. No door works well in an irregularly-framed opening. This rule holds true whether you're framing in a new wall or reframing an old wall.

With the full-length studs established or exposed on the outside of the door's proposed trimmer-stud location (see information on wall framing, page 40), the next step is to prepare the header. You can either use a 4 by 4 or make the header by nailing the widest dimension of two standard 2 by 4's together with a ½-inch plywood spacer in between. Stagger 12d or 16d nails along its length. Figure its length to be the distance between the two existing full-length studs nearest to the outside of each trimmer.

Cut away the sole plate as shown on the facing page. Watch out for nails and avoid cutting into the finish floor.

After measuring the exact width of the header (2 by 4's must be on edge for headers you make), figure the length of trimmer studs. They extend from the floor to the base of the header.

Before nailing them, however, prop and tack one trimmer against the inside of each full-length stud. Nail header in place (4 nails at each end), then toenail the cut-off full-length studs solidly into the top of the header.

Nail each of the outer trimmers to its full-length stud with 12d nails staggered along it's length. You'll probably have to block a third trimmer out from one side as shown in sketch to adjust for the proper door frame width, remembering to allow ½-inch on each side for shimming. Nail that trimmer into the block; toenail it into the sole plate and header.

How to Hang a Door

Hanging a prehung door frame within a rough stud opening is relatively easy. The main thing to remember is to keep the door frame plumb and level at all times.

As previously discussed, the stud opening is slightly larger than the size of the prehung frame. This is to allow for shimming the frame into exact plumb and level. To shim, the carpenter drives a pair of shingles together from each side of the frame between trimmer studs and the jambs to form a tight rectangular wedge. After framing is nailed, shims are broken or cut off flush with the trimmer studs.

Because the order of shimming and fastening is critical, follow this procedure:

1) Center frame in opening from side to side and from back to front.

2) Beginning next to lower hinge location, shim to estimated side clearance and fasten where stop molding will cover with two 10d finishing nails.

3) Insert shingles next to upper hinge location and tap them together until that side of the jamb is plumb, then nail where door stop will cover. Because shingles will compress slightly when you nail, you should allow a little extra shim.

4) Shim, check plumb, and nail halfway between top and bottom shims, where molding will cover.

5) Fasten door into position with hinge pins.

6) Shim and nail latch side of door frame at adjacent locations, checking to keep 1/16-inch clearance between door edge and frame. An

exterior door's sill and threshold should be nailed to joists at this time. The top of the sill should be flush with the finished floor level (see sketch on previous page). Attach with 10d finishing nails.

The door may not have a latch. A template for installing the door latch should be in the latch kit you buy. Plan the knob 36 to 38 inches above the floor. You'll need a brace, an auger bit, and an expansion bit to drill for latch insertion. Follow manufacturer's directions.

After installing latch, close the door and mark the top and bottom of latch where it contacts frame. Position the latch's striker plate and cut out the mortise for the latch.

Once the latch is installed, you're ready for the finishing touches: door stop molding and casing. Beginning at the hinge side, nail a length of door stop from the floor to the top inside corner of the frame with one 4d finishing nail every 12 inches, leaving 1/16-inch from door face on the hinge side so that the door won't bind in case you plan to paint it. Miter corners where stops join. Cut a bevel where stops meet floor to eliminate dust pocket problems and make the floor easier to clean around the base of the frame. Nail other stops flush with the face of the closed door.

Now nail the casing trim around the opening to both the trimmer studs and the frame's edges with 6d or 8d finishing or casing nails. Space nails 16 inches apart, set them below the surface, and fill the holes. When placing the molding, leave ½-inch margin on all jamb edges. Miter shaped moldings at top corners; butt-join flat, rectangular moldings.

For extensive wall repairs, see information on paneling, page 51.

Installing a Sliding Glass Door

Adding sliding glass doors to open a room to the garden or to provide a view is one of the most common remodeling projects in homes today.

Since the entire door and frame may be ordered from most manufacturers as one complete unit, installation is generally easier than it may look. Just follow the framing diagram and see page 48 if you're building the wall frame from scratch, and read the instructions under "Hanging a Sliding Glass Door." If you'll be putting the new sliding glass door in an old wall, follow the instructions in the previous section for removing the wall's covering, then cut away unwanted studs (as described in the preceding section on installing a door) and install the framing as explained below.

Ordering the Door

Sliding metal doors are available in a wide range of price and quality. Most types roll on bottom rollers, but there are also top-roller models. One type especially made for mounting on the outside of the wall rather than inside is particularly adaptable to remodeling jobs.

Choose steel if you want to paint your doors or aluminum if you want a door that won't require painting and prefer a metallic finish.

Door prices often include screens but not glass. The cost of glass is usually quoted as "extra" and varies in price according to type.

You may order your door from glass companies, builders' supply stores, or, in some cases, directly from the manufacturer.

Removing Wall Framing

Once the wall's surface is removed on both sides, as described on page 62, you are ready to remove studs, bracing, and any existing door or window jambs and frames.

One very helpful tool for this purpose is a reciprocating power saw. You can use it to cut studs or, with a nail-cutting blade, to cut through framing nails. If you don't own such a saw, you might consider renting one from your local tool supply.

If the wall is a bearing wall (as most exterior walls are), be sure to install some supporting device under the joists, parallel to the wall section being removed. The sketch at right shows one way to provide necessary support. (Add support beneath the floor if needed.)

To protect the ceiling finish, use a 12-inch plank and a blanket or several towels as cushioning. To set up the bracing support, have someone hold the plank against the ceiling while you force the 2 by 4-inch vertical supports in place. Protect a good floor finish with a plank or a piece of plywood.

Before you saw into the studs, measure carefully from the subflooring, marking each stud at line A as shown in sketch.

Saw through the studs as described under "How to Install a Door"; then pull the studs off the frame. (Sometimes you can plan to drop a whole section of wall with inside plaster intact if you've already removed your exterior wall surface.)

Next, saw through and remove the section of sole plate where your door is to be set. The opening should

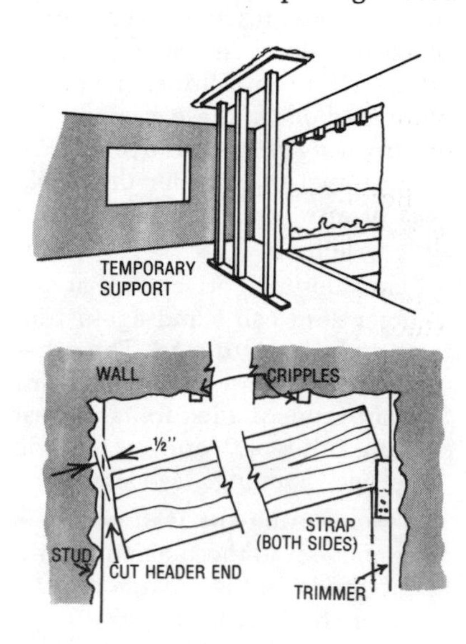

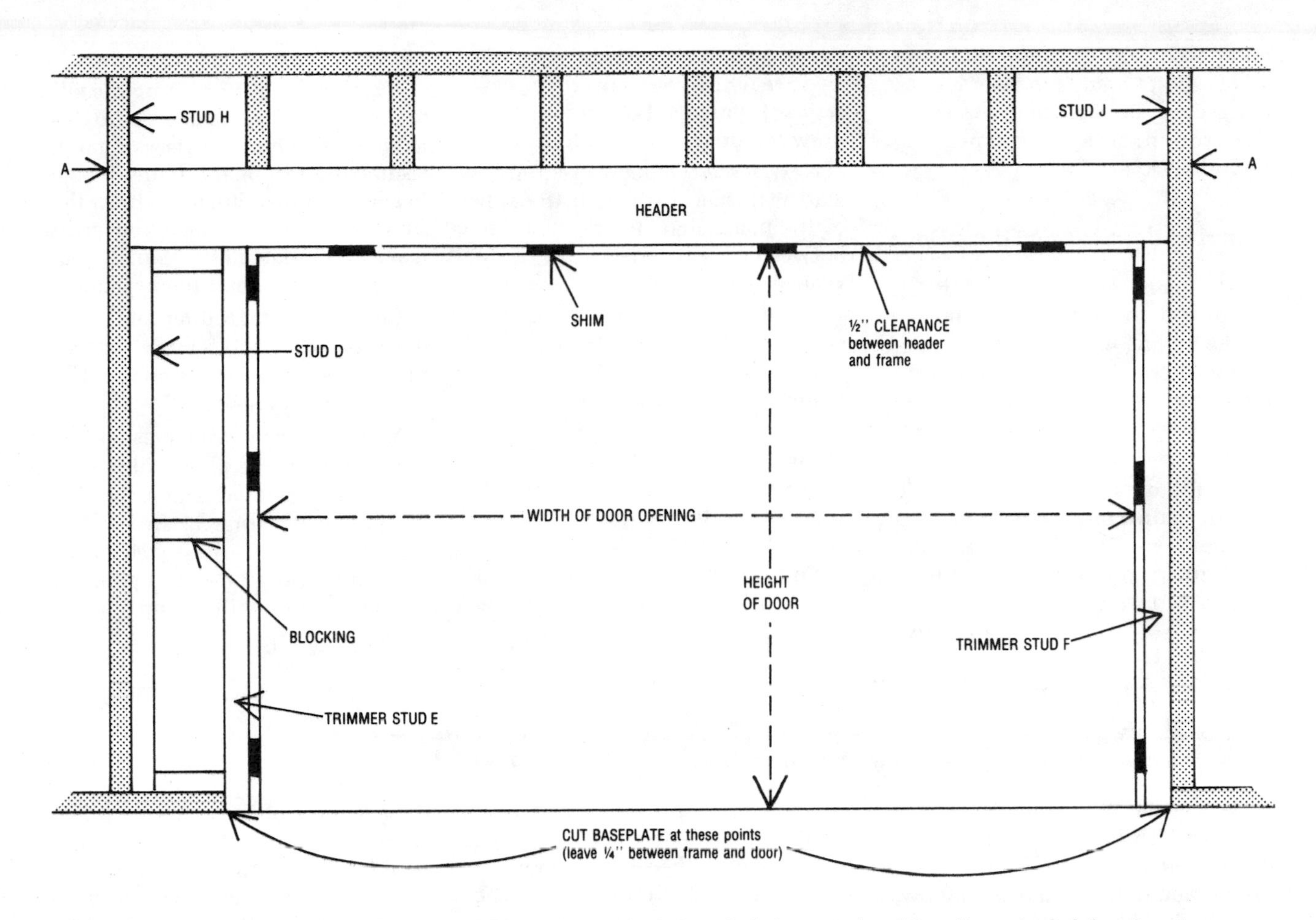

SLIDING GLASS DOOR is framed in an wall as shown. Shaded studs belong to the old wall; unshaded framing is new.

allow ¼-inch leeway on each side of the door (see sketch above).

If you have had to remove an existing door frame, you may have to block in the opening left in the floor to bring it level with the subfloor. Since it must support a portion of the sliding door's weight, the subfloor must be level at all points and have a secure bearing.

Before continuing, recheck the actual depth of the header delivered to you from the lumberyard and figure exactly how long the trimmer studs (D), (E), and (F) will have to be cut to fit tightly. (You should be able to use the studs removed from the wall for these trimmers.) Check the required length of header as well. It, too, should fit tightly between the existing studs (H) and (J). Trim it if required and cut as shown in the sketch above to avoid binding when you set it in place.

Set trimmer stud (F) in place and nail it securely to stud (J). This gives you initial support for your header so that you can hoist it into place.

Line up your header with the opening and hoist one end so that it rests firmly on the trimmer stud. If a long header has any bow or warpage to it, install it with its crown facing up.

Anchor the header as shown in the sketch; then get the second trimmer stud ready to install but don't nail it in place.

Hoist the header into position and support it while you force the trimmer stud underneath, flush with the existing stud. One expert installer often uses an automobile jack to force a particularly troublesome beam into place.

As soon as you have the header in place, nail the trimmer stud to the existing stud and toenail the header at both ends.

Next, check to see that all the short cripple studs above the header are in direct contact with it. Force shingle wedges into spaces between cripple studs and header. You should also check the subfloor to make certain that it is absolutely level. Cut down any humps with a plane and shim depressions with pieces of shingles after the frame is in place.

Your next step (unless you live in a concrete slab-floor house with outside paving level with inside floor) is to install the flashing. Most manufacturers do not include flashing with a door unit. Check and order separately if necessary. Butt the vertical flange of the flashing to the subfloor edge; then nail or screw it in place.

Apply a bead of mastic along the length of the flashing as recommended by the manufacturer and across the two ends to form a dam. On a concrete slab that sits above the ground, this is an especially important job.

Hanging the Sliding Glass Door

Now you are finally ready to install the unit door and frame—another two-man job. If you buy an aluminum door that has to be assembled, be sure that you have assembly instructions. Steel doors are always delivered already-assembled.

Set the door frame in place and snug it against the trimmer stud that will give you the desired opening location (refer to sketch). Insert several screws loosely through the frame into the trimmer stud and also the header beam to hold the door in place while you make the necessary adjustments. Be careful not to draw the door frame up to header.

Next, force the opposite trimmer stud in place, flush with the side of the metal door frame. Block in between this trimmer stud (E on sketch) and stud D.

Determine which side of the opening is the locking side and force two shingles as shims between the frame and trimmer stud on that side, as described in the previous section on hanging a standard door. These shims leave a margin of space so that the locking mechanism won't be jammed if the stud warps.

Once you have blocked and shimmed as necessary to true up the door frame, tighten all screws holding frame to studs.

Install screws, tightening the sill to the subfloor, again using shingles as wedges where necessary. Finally, secure the top of the frame to the header, being careful not to draw the frame out of line, since you must leave approximately ½-inch between the frame and the header.

Check doors for rolling, and, if the door is adjustable, make adjustments as recommended by the door's manufacturer.

Your only remaining jobs will be installation of the glass and the addition of the interior and exterior moldings. Patching of inside and outside wall finishes (see walls chapter; pages 40–56) and painting the new work will round off the project.

Installing a Movable Window

Today, installing a movable window is much like installing a standard door or sliding glass door, except that in the case of a window, the opening normally stops short of the floor. Because movable windows are manufactured and sold in prehung units, like doors, the job is usually quite simple. All you do is prepare the rough stud framing to receive the window, slip the window in—casing and all—level it, and fasten it.

A wide variety of window styles is available. The type of window used largely determines exact installation procedures. For instance, wood casement windows are generally shimmed into proper level, as explained here, but most aluminum windows are simply propped into position, leveled, and nailed to outside of wall framing. Find the type that best suits your needs and bring it to the building site before beginning work so you can use it to make precise measurements as you work.

You'll be working either with a new wall, where you construct the window's rough framing from scratch, or an old wall that you'll need to open up and reframe. If you'll be reframing, use the same methods described in the previous two sections for cutting into the wall. Of course, measuring the opening will differ.

Locating the Opening

The amount you allow for adjusting the window between the header and sill will depend on the type of window you buy. Aluminum windows don't need to be shimmed, but since these windows don't have a finished sill, you must allow extra space between the window's base and the rough sill for a finish wall covering. Extra space also may be needed for operating a crank.

For complete wood casement windows, allow ⅜-inch at top, bottom, and sides.

Most windows are placed in walls so their top line is 6 feet 8 inches above the floor. This way, top lines of windows and doors will generally match.

To figure where to cut into the wall, add the header's exact width, plus allowance for shimming window at top and bottom, plus 6 feet 8 inches. Measure the sum up the wall from the floor, making a mark at each side of the window's approximate location. By drawing a line between these two marks, you get the important top line for cutting into the wall and the location where the header's top will meet the bottom ends of the cut-off wall studs.

Below a mark at 6 feet 8 inches, add the window's height, plus shimming allowance top and bottom to the width of the sill: now you have the location where you'll cut the lower portions of the existing wall studs—the line where sill will rest on the cut-off studs.

Locate the nearest existing outer wall stud to each side of the window's proposed location. Open the wall covering to those studs. Before cutting intermediate wall studs, it may be necessary to provide support for the ceiling (see page 66).

After opening the wall, use a square to carefully mark each wall stud within the proposed window location at both the top and bottom marks where they'll butt against header and sill. Cut the studs squarely with a saw.

A Window's Proper Framing

With wall framing ready to begin constructing the rough opening, take a close look at the sketch.

You'll probably need a helper to aid in lifting the header and prehung window into place. Position a large header, using the same methods described in the section on framing a sliding glass door. If the header is small, set it in place as you would for a standard door.

Nail trimmers into place, staggering 12d or 16d nails along their lengths. Toenail cut-off studs to header's top and toenail header's ends to full-length studs.

If you're going to adjust the opening's height or level at all, better do it before nailing the double sill in place. Be sure the sill is level. Nail the bottom 2 by 4 of the double sill to cut-off stud ends, and add an extra cripple stud at each end. Nail the top 2 by 4 to the bottom 2 by 4 and to the full-length studs at each end.

The last step of rough framing is to nail double studs to each side of the opening between header and sill. Allow ⅜-inch clearance on each side for shimming the window into exact level.

Hanging and Finishing Window

Once the rough framing is complete, hang the window according to manufacturer's recommendations. This process should be similar to methods used for hanging a standard or sliding glass door discussed earlier. For maximum protection against weather, tack building paper around the outer perimeter of the window.

Apply moldings (page 27) and finish wall coverings (see information on applying sidings to exterior walls, page 42, and paneling an interior wall, page 51).

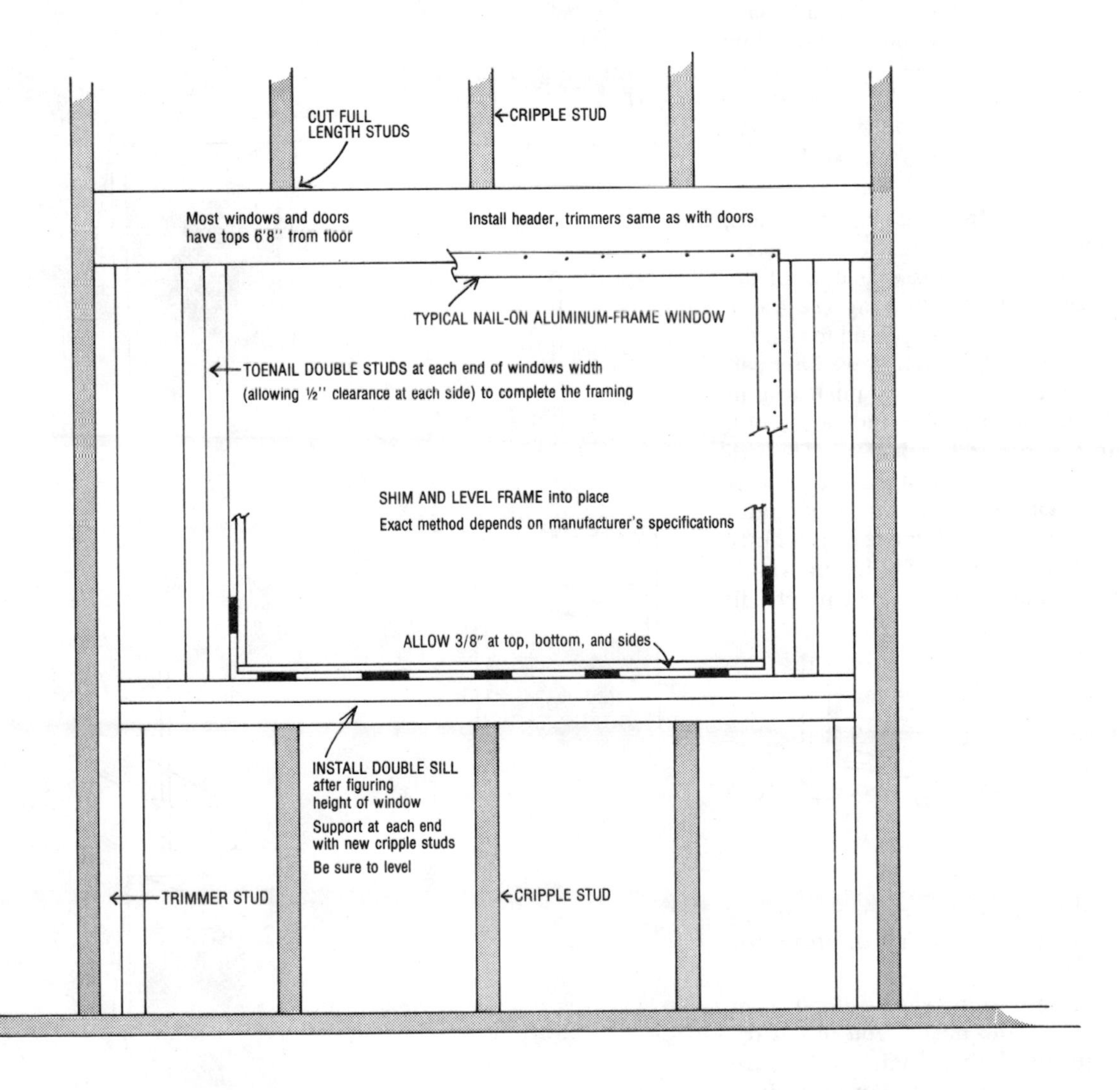

Adding a Simple Stationary Window

Not every window needs to be opened for air. Some windows—one next to a front door or those high along a wall—are added just to let light in or to expand the view from inside. Installing a stationary window is considerably easier and less expensive than adding a movable prehung window: easier because the glass dealer will cut the glass to fit a frame you first build, and cheaper because you need not pay for the prehung frame—just for glass and standard lumber.

Once you have the rough stud framing prepared for a window (as described in the preceding section on movable windows), you'll have to build a frame to hold the window glass in place. The designs on this page should work for windows of most sizes, but if your window will be large, check out your exact design with the glass dealer.

Build the frame as shown, making sure you keep the frame square. Then have the glass dealer cut the glass to fit. Caulk along the frame with a glazing compound to seal the joint around the window. After setting the glass in place, push around its edges so compound oozes out, add another bead of compound around perimeter of glass, and carefully nail a "stop" into place with finishing nails. Remove excess putty with a putty knife. Using small triangular metal "glaziers points" to

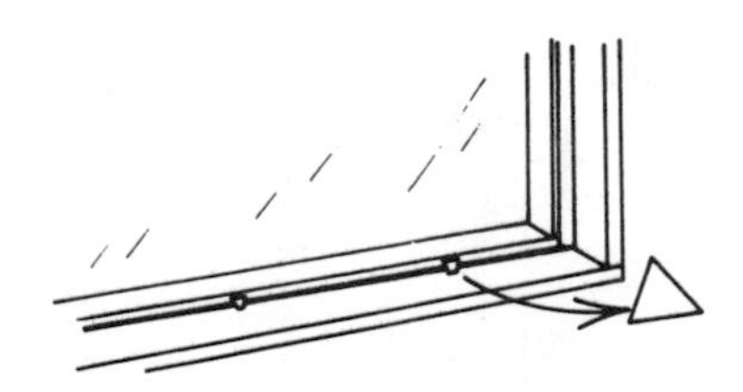

hold the glass in position, you can apply putty instead of stops for small windows.

If the designs shown don't meet your specific needs, you might try adapting their principles. Be sure that glass is well supported around all edges.

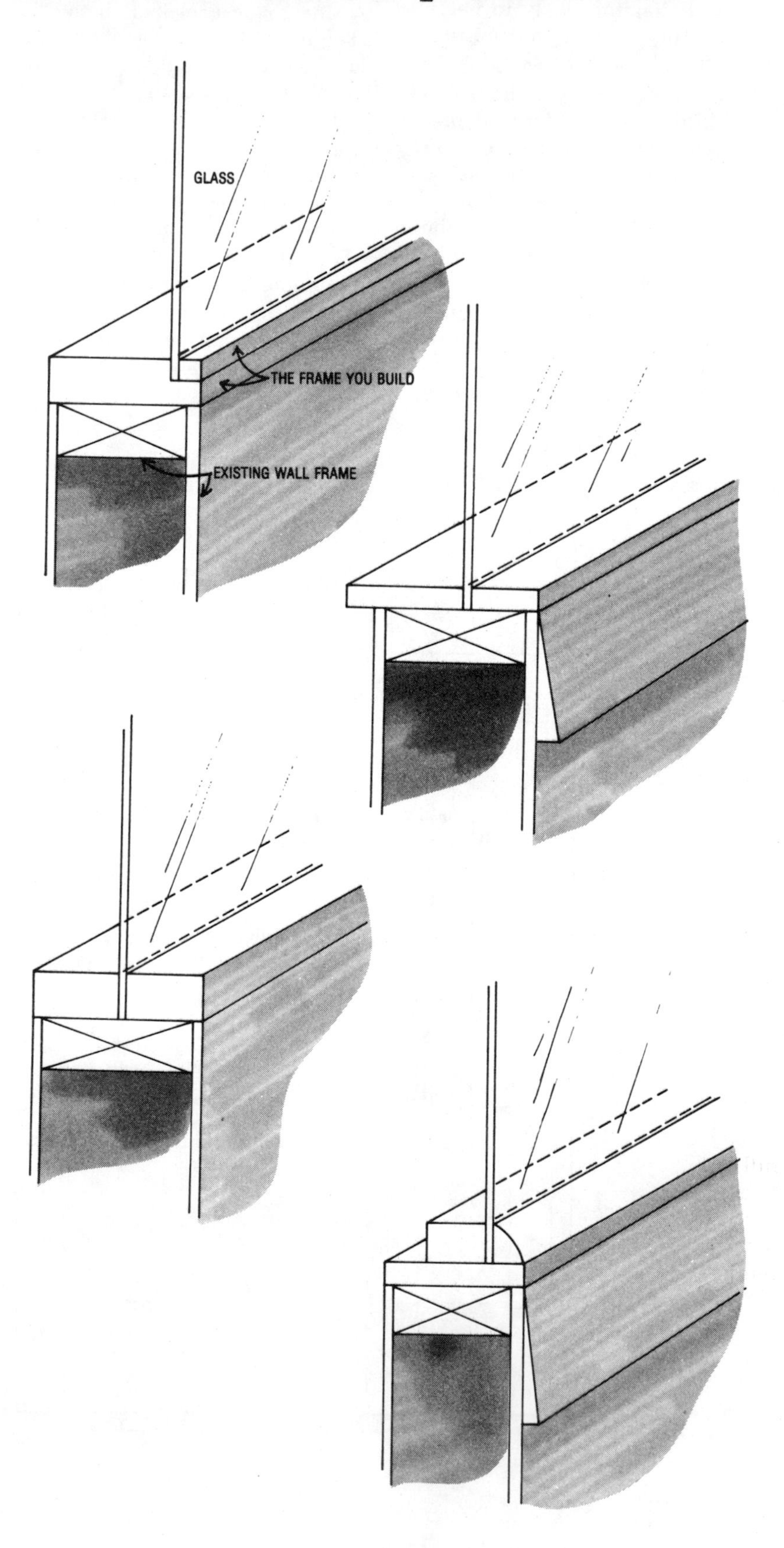

How to Install a Skylight

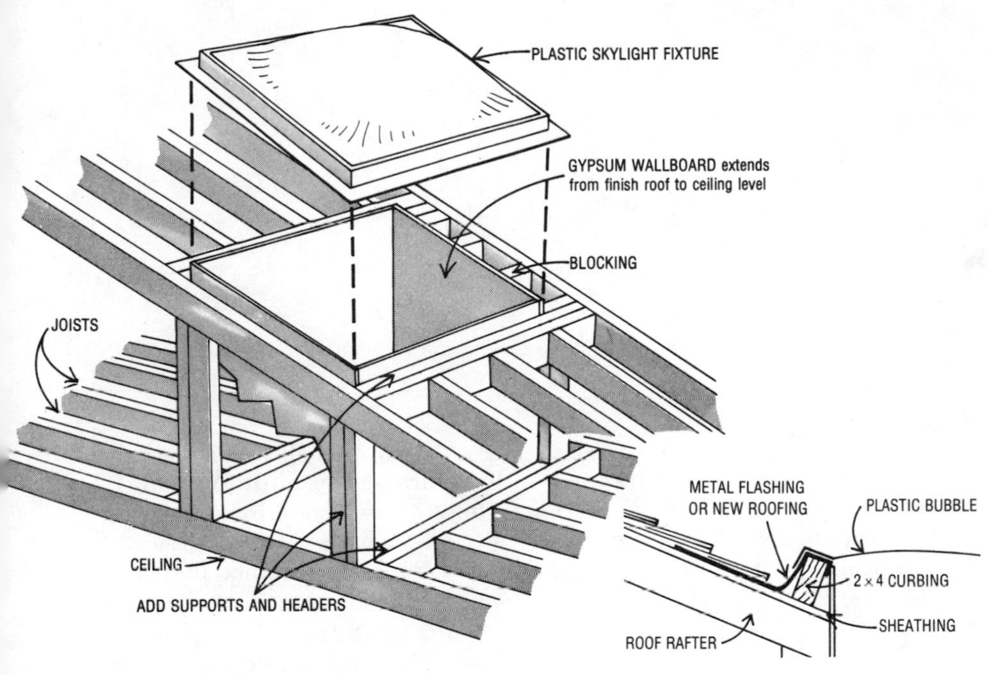

Skylights have traditionally been highly valued by artists, mostly because they are a shadowless source of light. But even if you're not an artist, a skylight is the best and easiest device for bringing natural light into the interior of a house where normal windows are absent or far away.

Several different styles of preformed acrylic plastic skylights are available at glass and window dealers or directly from skylight manufacturers. Made for easy installation by the do-it-yourselfer, they come in rounded, rectangular, and square shapes, fitting both flat and pitched roofs.

As you can see in the sketch, a full skylight unit usually consists of two main elements: the plastic "window" body you buy and attach to the roof surface, and a wallboard light shaft that you build between roof and ceiling.

Although skylight installation is not difficult, you must carefully seal the opening in the roof and properly reframe openings cut in joists and rafters. Work of this type is always governed by local codes, so be sure to check regulations before beginning. It's possible that you may need a building permit.

If you install the skylight in conjunction with a reroofing job, the work is simplified; however, the instructions that follow assume using existing roofing with a minimum of patching. Resealing is no major problem with composition shingles or built-up roofing; if you have wood shingles or shakes, take care, especially along the top side, to prevent leaks.

1) Begin with the light shaft. Since the size and shape of the ceiling opening will vary with the slope of your roof, you must determine the measurements of the light shaft from the top down.

First, check the placement of the ceiling joists and roof rafters in the attic or crawl space. You may be able to avoid cutting extra ones by shifting the opening slightly to one side.

2) Locate the two upper corners for the roof opening, then drive a long nail up through the roof sheathing at each point. Take your skylight out onto the roof, set it over these nails, and locate the two bottom corners. Drive nails down through the roofing at each point, then go back inside and use a plumb bob to align the ceiling opening with all four corners of the roof opening. When doing the rough framing, remember to allow an extra ½ inch on all sides of the light shaft for gypsum board, if you plan to use it.

3) Because most skylights are made to extend across the area of two or three joists, you will have to cut away sections of ceiling joist and roof rafters and reframe with headers and blocking as shown in the diagram. It is possible to do much of this work from inside while the roof is still closed.

4) Before cutting the opening in the roof sheathing, mark a line on the roofing far enough back to allow for the raised frame or curbing around the opening. Carefully cut and remove the roofing to this line. Then *before* attaching the curbing, slip metal flashing well up under the shingles or shakes on the top and sides around the hole. It may be necessary to cut or pull some nails in order to get the flashing far enough in.

In flashing down the sides of a skylight in a shingle roof, work separate metal plates into each course, rather than using a single continuous flashing. Otherwise, water might run in sideways under the shingles. Allow enough flashing so it can be bent up and turned over the curbing after the curbing has been nailed in place.

5) Install curbing. Bend flashing up against sides and over top of curbing. Seal any gaps or joints in flashing with roofing mastic.

6) Follow the manufacturer's instructions for attaching the plastic skylight. Caulking may be needed to fill voids and cover nails.

7) Add ½-inch gypsum wallboard to the shaft, extending from ceiling level to finished roof level. For information on working with and nailing gypsum wallboard, see page 52. If your ceiling is wallboard, cover the corners formed where the four walls of the light shaft meet the ceiling.

8) For finishing touches, paint the shaft white. A translucent light diffuser can be added to the ceiling.

Floors

On some parts of the globe, people still walk on dirt floors. In most areas of the United States, though, homeowners are fortunate enough to have finished floors that are raised well above the ground.

This chapter discusses the framing of the floors we walk on and gives methods for laying wood flooring. A brief description of methods for building simple stairs and an illustrated guide to building a deck are also given.

FLOOR FRAMING

Floors of houses, unless formed of concrete like basement floors, are constructed from several different structural members. Typical members making up floors include sills, posts, girders, joists, bridging, and subflooring.

All of these elements are not present in every floor's framing. A masonry wall usually supports girders on the lowest story; posts may substitute if the basement has full headroom. On stories above the first, a wall usually supports joists instead of posts or girders. Of course, a concrete slab floor has none of these elements.

Below is basic information about each flooring member.

Sill Construction

The system of joining framing members to the foundation and to each other ("sill construction") depends upon the type of house framing used—balloon or western (page 41). Differences are shown in the two small sketches below.

Posts

Sometimes used intermediately to support the lengths of first-story girders, posts have sizes and spacings strictly controlled by codes. Typically, their maximum spacing is 8 to 10 feet, their minimum dimensions 4 by 4 inches.

Girders

Although usually solid, wooden girders are often built-up from two lengths of 2-inch-thick dimension lumber spiked together. Their sizes, like the sizes of all framing members, are set by building codes. These codes determine a girder's size on the basis of the material it's made from, the distance it spans, and the load it bears.

Ends of girders bear solidly on foundation walls. Sometimes they are set into a slot provided in the foundation so that the girder's top is level with the top of the sill.

Although builders also use steel I-beam girders, these are characteristically outside the realm of the basic carpenter.

Wood Joists

Joists distribute a floor's load to the girders, bearing walls, or other main-bearing supports. Made from 2-inch-thick lumber, joists vary in

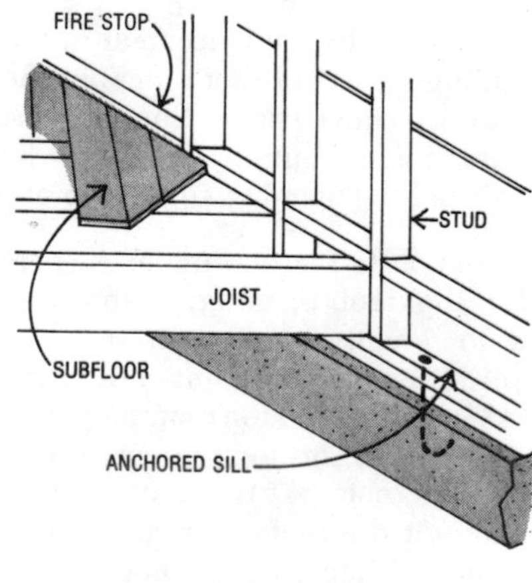

BALLOON FRAMING

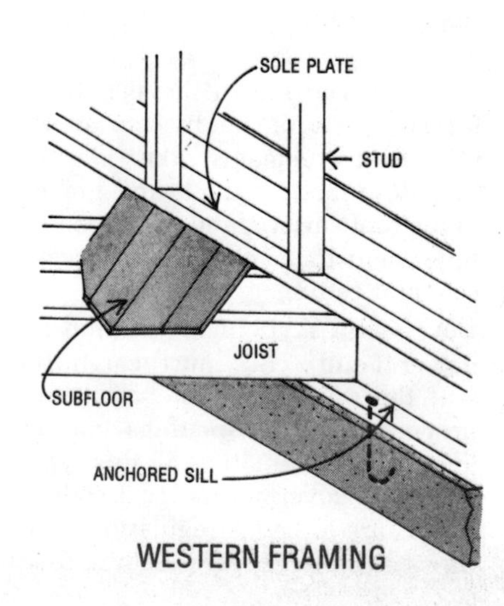

WESTERN FRAMING

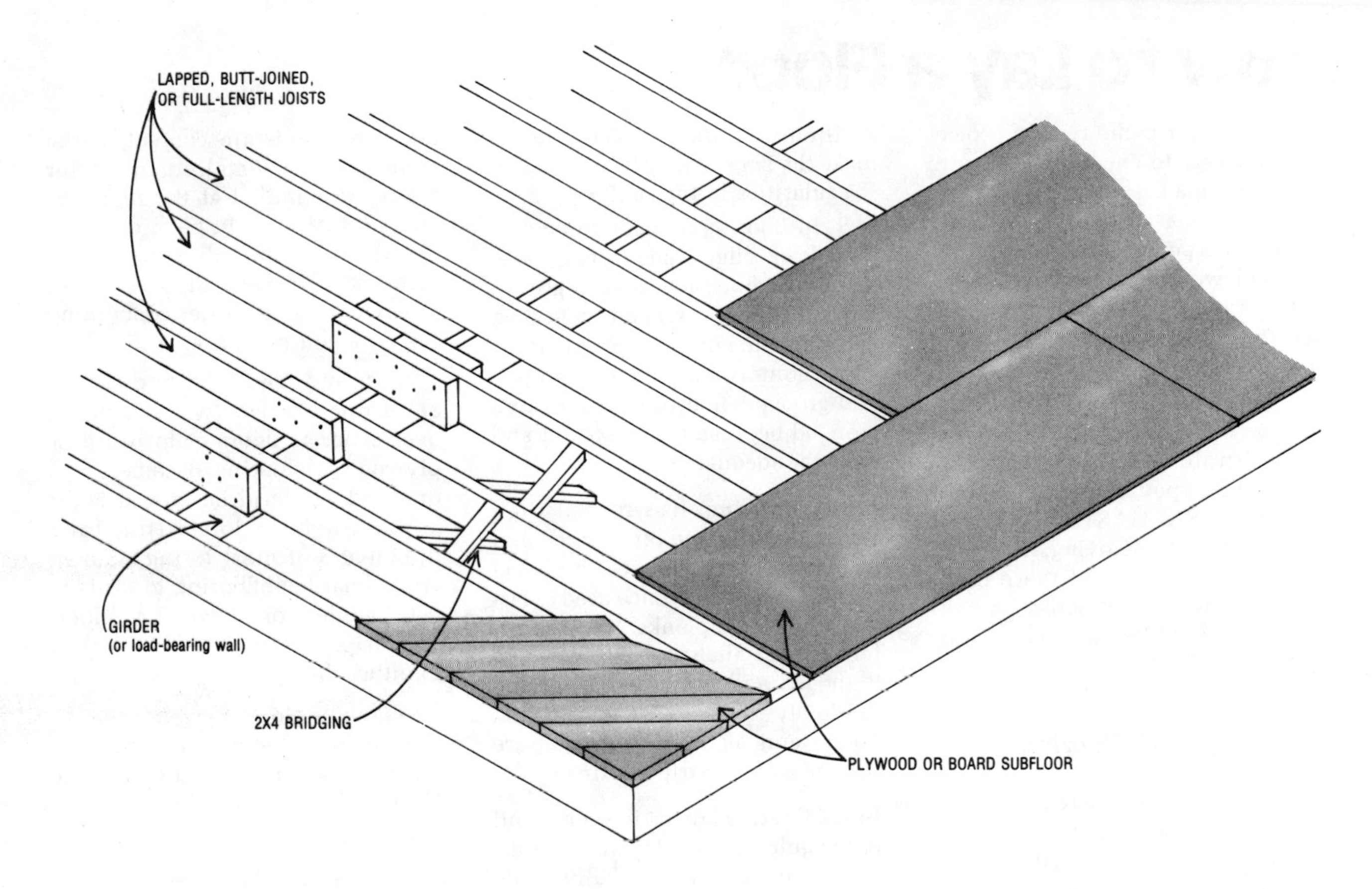

width—6, 8, 10, or 12 inches —depending upon the load they carry, the length they span, the spacing between them, their lumber species, and their lumber grade. Local building codes determine exact sizes and spacings used for specific loads and spans.

Standard spacing of joists is 16 inches on center, but sometimes they are spaced 20 or 24 inches on center.

The wood species used for joists should be strong, stiff (to minimize vibrations and plaster cracking under live loads), free from warp, and able to hold nails firmly. Though most softwood framing lumbers are acceptable, be sure to check with codes governing framing size and grade.

Bridging

Bridging helps to stiffen joists and transfer loads concentrated on single sets of joists. It is fastened between joists at the approximate middle of any span longer than 8 feet. Cross bracing (shown above) is commonly used. An easier method is to cut short blocks from joist lumber and nail them between joists like staggered fire blocking (page 48).

Subflooring

Applied on top of joists and beneath the finish flooring, subflooring is of two general types: board or plywood.

Board subflooring may be square-edged, shiplap, regular tongue-and-groove, or end-matched tongue-and-groove. Boards are 1 inch thick by 4, 6, or 8 inches wide. Depending upon the species, Number 3 or Number 4 Common are the minimum lumber grades used.

Subflooring boards can be applied either diagonally or perpendicularly to joists. Their direction determines the direction wood finish flooring can be applied. Over diagonally-applied subflooring, it can be applied either horizontally or perpendicularly to joists, but it must be laid at right angles to subflooring that is applied perpendicularly to joists.

Subflooring boards always end over joists unless end-matched tongue-and-groove boards are utilized. In the latter case, stagger joints so that no two successive boards are joined in the same space.

Plywood subflooring is available in several sizes. The size chosen for particular applications is determined by the joist spacing and the type and direction of finish flooring being laid on top.

For suggestions of proper subflooring thicknesses and joist spacings, check with your local office of the Federal Housing Administration, and be sure to check local building codes. For any finish flooring material other than finish wood strip flooring, install solid blocking between joists under all plywood edges.

How to Lay a Floor

Laying a floor is the type of project you're likely to encounter if you're remodeling a basement or a garage. Only certain aspects of this job concern the carpenter—laying subflooring and working with wood flooring. For information on subflooring, see the previous section on floor framing.

Finish flooring materials other than wood (composition tiles, vinyl tiles, linoleum, rubber, ceramic tiles, and carpeting) are not discussed in BASIC CARPENTRY. For information on removing, repairing, and replacing some of these products, see the Sunset books BASIC HOME REPAIRS, or REMODELING WITH TILE.

Types of Wood Flooring

Wood flooring is generally thought of as being "hardwood flooring" or "oak flooring." Actually, wood flooring may be of either hardwood or softwood. Other typically used hardwoods include maple, beech, and birch, and other softwoods used are fir and pine.

Wood flooring is milled in three general styles: strip flooring, plank flooring, and block flooring. Each of the three comes in varying lengths, widths, and thicknesses.

Strip flooring, often found in formal settings, is basically long and thin. Lengths vary and are applied at random. Thicknesses are $^{5}/_{16}$, $^{1}/_{2}$, $^{25}/_{32}$-inch; widths vary from $1^{1}/_{2}$ to $3^{1}/_{4}$ inches. The underside of each strip is usually hollowed out for better resiliency underfoot. This helps make the floor fit tightly despite any irregularities in the subfloor.

Strip flooring is made to be butt joined at edges and ends, face-nailed or joined tongue-and-groove fashion at edges and ends, and then blind nailed. For areas experiencing sharp contrasts in climate, tongue-and-groove flooring is recommended because it can expand and contract adequately.

Plank flooring, less formal than strips, is wider and often bored and plugged at ends to give the effect of the wooden pegs once used to fasten down such planks. Widths, differing by 1 inch, range from 3 to 9 inches and are usually applied randomly like the strips. Thicknesses and methods of joining are the same as for strip flooring.

Block flooring comes in square and rectangular units, often in the familiar "parquet" style. Usually made from three or four short strips of flooring glued on mesh or joined with splines, blocks are from 5/16 to ¾ inch thick. The square or rectangular dimensions commonly range from 4 to 16 inches on a side.

Flooring grades are established according to the wood's color, grain patterns, and defects.

General Application Notes

Some types of wood flooring should be delivered to the building site a week or so early, allowing them to attain the moisture content of the room they're installed in. Other types are bundled at the moisture content they should be laid at and should be installed immediately after bundles are broken open. Check with your dealer concerning the type you purchase.

Strip and plank flooring can be applied to sleepers (wooden strips) over concrete floors or to board or plywood subflooring. Because strips or planks should be laid at right angles to floor joists (the long dimension of most living rooms), apply board subflooring to joists at a 45° angle. For plywood subflooring, strips or planks can be applied in either direction.

Though block flooring can be laid on any type of wood subfloor, only laminated wood blocks should be installed on concrete.

Strips or Planks on Subflooring

Before beginning application, check the subfloor to make sure it is clean and well nailed to the joists to prevent it from squeaking later. By laying a heavy building paper or deadening felt, you can slightly lessen noise and cold.

If the room you'll be flooring is fairly narrow, you can start by installing the first strip or plank at one wall. If the room is quite wide, however, you should begin in the center of the room and work each way.

To start at a wall, space a long,

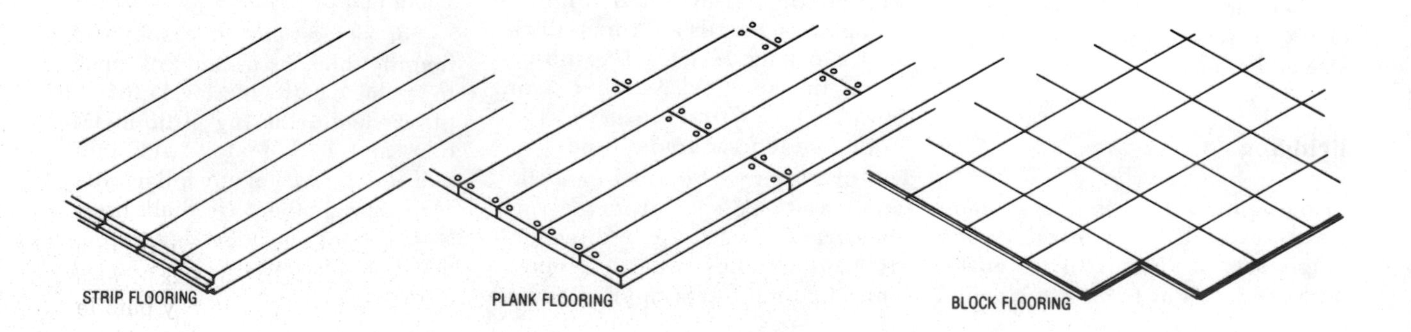

BELOW are methods of nailing square-edged and tongue-and-groove flooring.

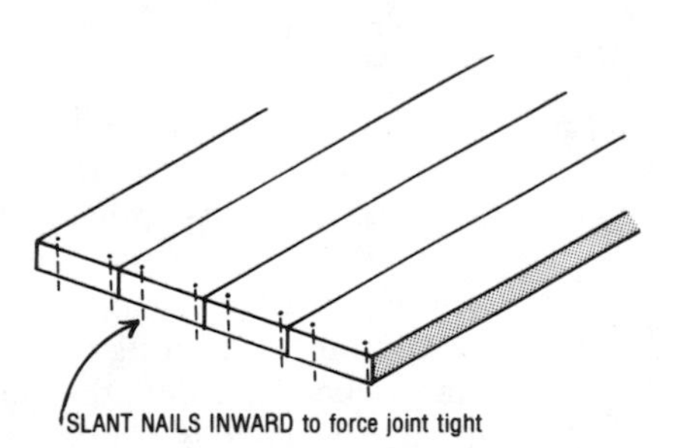

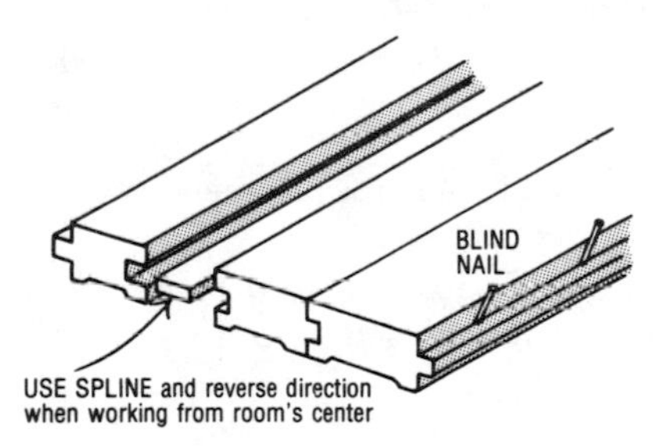

FROM A BIRDSEYE VIEW, illustration at right shows proper method of laying strip or plank flooring from end of a room.

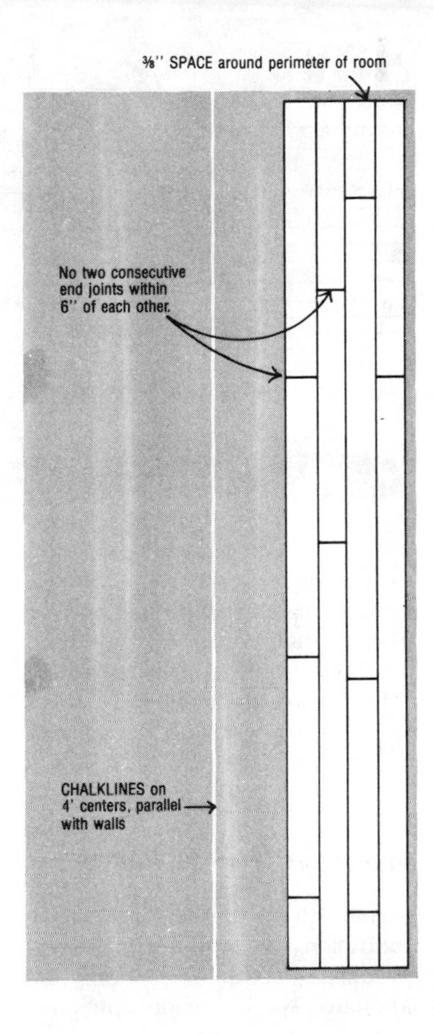

straight strip of flooring about ⅜-inch from the wall (leave this space around the room's entire perimeter). Adjust the strip so it sits exactly parallel to the wall. Keeping the flooring parallel to both walls as you work across the room is very important. Once you're sure that a piece is square and straight, nail it to the subfloor and joists. If face nailed, countersink nailheads; fill them later.

To start in the room's middle, carefully measure from each wall to the room's center. Snap a chalk line parallel to and equidistant from both walls. Lay a long strip of the flooring along this line and face-nail it in place, using the same methods described above. For tongue-and-groove flooring, cut a hardwood spline to fit as shown, connecting the two grooves meeting on one side of the flooring strip or plank.

To continue the job, no matter where you started, lay out a few strips or planks, arranging them so that no two consecutive end joints fall within 6 inches of each other and so that the wood's grain and color match attractively. Try to keep all grain running in the same direction. To guide you in keeping the flooring lengths straight and parallel to the walls, snap chalk lines about every 4 feet from end to end across the room.

Nailing will depend on whether you're applying tongue-and-groove or square-edged flooring.

For tongue-and-groove, blind-nail with one nail every 10 to 15 inches. Nail sizes depend on the thickness of flooring you choose. Use 8d for 25/32-inch flooring, 6d for ½-inch, and 4d for ⅜-inch. Choose only flooring nails or nails with ringed or annular grooves (see page 29). Nails should penetrate through subfloor into joists, especially where plywood subflooring is used.

For square-edged flooring, face-nail across each strip or plank's width every 8 to 10 inches. The number of nails at each nailing location depends on the flooring's width. Use two nails no closer than ½-inch from each edge for 2-inch flooring, four for 4-inch flooring, six for 6-inch, and so forth.

Nail carefully to prevent the floor from squeaking later and to keep from marring the floor's finished surface. Use a nailset to set nails below the wood's surface, being careful not to smash any part of the flooring.

Professional floor installers use mallet or power-driven nailers that drive nails rapidly and accurately. Inquire about renting one of these nailers from a local tool supply company to speed up and simplify your flooring work.

After installing five or six rows, fit a piece of scrap flooring against the outer row and give it a few sharp raps with the hammer to press all the members tightly together. Do this every five or six rows along the flooring's length. Be sure all lengths stay in alignment with chalk lines.

Stop about 2 or 3 feet short of the far wall. At this point, check to make sure the rows are still truly parallel

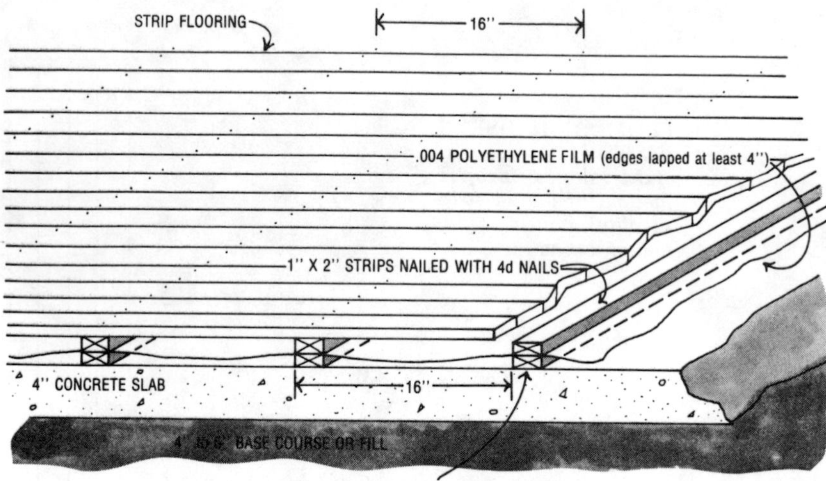

to that wall. If they're not, slightly plane down edges at one end of several remaining strips to adjust. When planing tongue and groove, be sure to plane both the grooved edge and the adjoining tongue.

Applying Blocks to Subfloors

Remember—though wood block flooring may be laid on either wood or concrete subfloors, use only the laminated wood type on concrete.

Lay out the block's locations carefully, snapping chalk lines on the subfloor to keep the rows straight and parallel with walls. For professional-looking fitting, adapt the methods used for planning ceiling-panel placement as given on page 59. Blind-nail the blocks through the subflooring into joists (just as you would in strip flooring) or apply with special adhesive. Be sure to butt the blocks snugly together.

Strips or Planks on Concrete

You can apply any of the three finish flooring styles over a concrete-slab floor after applying a moisture barrier. The main difference between block flooring and strip and plank flooring is that you can't apply strips and planks directly to concrete. For this you need a nailing base, usually consisting of 2 by 4 strips (sleepers) fastened to the concrete.

One way to create a moisture barrier over a concrete slab is to treat the slab with an asphalt primer, "hot-mop" it with a coat of asphalt, spread a layer of 15-pound asphalt-saturated building paper, and apply another coat of hot asphalt. As you might guess, this is not an easy job.

Another rigorous method is applying a coat of asphalt primer, a coat of asphalt, a layer of polyethylene film, and a finishing coat of special asphalt mastic in which sleepers may be set.

Some latex-type sealers are also available for sealing floors. You trowel them on in two parts. Check sealer manufacturer's directions for specific application information.

Easiest of all is the method shown in the illustration. Faster and less expensive to do, this method uses a double layer of 1 by 2-inch wood sleepers soundly nailed together with a layer of polyethylene in between.

The bottom 1 by 2's (pressure treated redwood) are secured to the slab on 16-inch centers with a latex mastic that's rapidly applied with a caulking gun and then fastened with concrete nails every 24 inches along their length.

Next, a moisture barrier of .004-inch (4 mil) polyethylene is laid across the lower 1 by 2's and held in place by the upper 1 by 2's. The last step is nailing 25/32-inch tongue-and-groove strip flooring at right angles to sleepers, one nail at each bearing point. For smaller sizes or square-edged flooring, first apply a layer of ¾-inch exterior plywood as a subfloor.

The method shown provides dead-air spaces above and below the polyethylene that prevent the rise of moisture and reduce the amount the floor will expand and contract under varying heating conditions. If any wood besides the pressure-treated sleeper is subjected to the moisture beneath the floor, that area should be vented to the outside to protect against dry rot.

Applying Blocks to Concrete

Laminated block flooring is ideal for laying on concrete floors. Two advantages of blocks: 1) they will ride over minor bumps and valleys as no long, straight flooring can do; 2) they can be applied with mastic directly to the floor without sleepers.

To apply blocks with mastic, you need a smooth, level, dry surface.

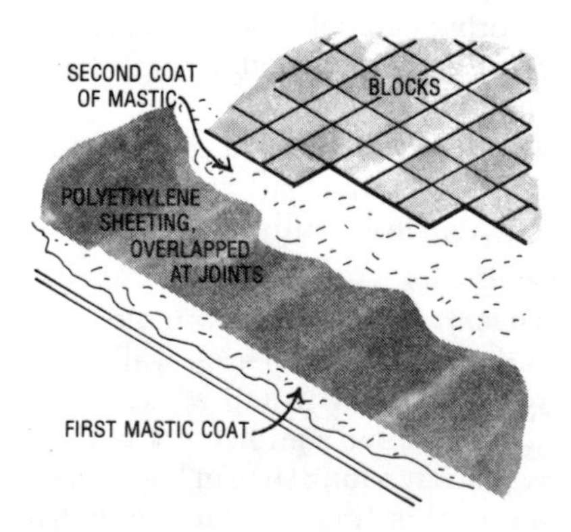

Apply a thin coat of mastic, lay 4-mil polyethylene sheeting on top, add a second coat of mastic, and set blocks in place. Since instructions may vary slightly with the manufacturer, read directions accompanying the mastic.

Planning the blocks' placement carefully before you begin work will pay off when you're laying the blocks in place. (See information above on laying blocks on subfloors.)

Building a Basic Stairway

Stairs are usually built by a craftsman during the construction stages of a house. The variety of types of stairs is virtually endless, ranging from a short basement stair to complex main staircases. Specifications and techniques used in building complex staircases are well outside the realm of basic carpentry. But a simple open stringer staircase—perhaps only three of four stairs high—leading to the basement or onto a deck can be constructed by most beginning home carpenters.

Shown in the illustration are the basic components of a simple open-stringer stair. The main supporting element is the "stringer." The boards you walk on are fittingly called "treads", and the piece rising behind each tread is called the "riser." The part of the tread that sometimes extends out over the riser to take wear is known as "nosing."

The approximate sizes of treads and risers are strictly governed by building codes so that stairs will be strong and easy to climb. You will have to calculate the exact size of these elements according to the precise height the stair will span. For purposes of calculation, assume that your stair will stretch from one finish floor to another. In reality, the stair may extend from a basement floor up onto a cement slab at ground level or from a dirt floor onto a deck.

The first procedure is to stand a straight piece of small-dimension lumber (a "story pole") in the gap that the stair will span and mark carefully where the top floor level crosses it. Measure the exact distance from finish floor to finish floor—this figure, the total stair rise, must then be divided into equal step rises. Because 7 inches is a good approximate riser height (check local codes), divide the total rise by 7 (or whatever measurement your building codes specify.)

A specific example may help to clarify the procedure. If the total rise is 36 inches, divide 36 by 7. The answer is $5^{1}/_{7}$. Because you can't have $5^{1}/_{7}$ risers, you must adjust the 7-inch figure slightly. Five risers come closer to hitting the 7-inch figure than 6 risers. Five risers would each have a $7^{1}/_{5}$-inch rise, six risers would each have a 6-inch rise.

Set $7^{1}/_{5}$ inches on a pair of wing dividers and step off this distance on the story pole from the top-floor mark to the bottom floor line. If the last step of the wing dividers doesn't end exactly on the bottom floor line, adjust them slightly and repeat until it does. When it comes out exactly, the dividers are set at the height of each riser.

Mark this distance across the tongue of a framing square with a grease pencil. Mark the width of the tread (about 9 inches—check codes) across the blade. Lay the square close to the "top" end of the stringer stock (use 2 × 10 or 2 × 12) as shown. Leave enough room to cut the board's end off at the angle formed by setting each mark on the square directly over the board's edge. Carefully mark the rise and run along the outside of the square with a sharp pencil. Next, slide the square along the board so that the mark on the tongue is directly over the end of the pencil mark just drawn along the blade and repeat (see sketch). Do this for each tread and riser on both stringers. Extend the last tread and riser lines at each end of the stringer to its back edge.

Once this layout is complete, carefully cut along all layout lines. Because the tread thickness will add to the first step height and diminish the height of the last step, measure the exact thickness of the tread, cutting this amount off of the bottom of the stringer.

For treads, use clear, dry, edge-grain 2 by 10s. For risers, use matching 1 by 8s or 1 by 10s; rip them to proper width.

Risers are nailed to stringers first, using 8d finishing nails. Then add treads, nailing them to stringers with 12d finishing. Using 8d nails, nail the bottom edges of risers into the back of treads. Wherever you place the finished stair, be sure to anchor it securely at both top and bottom.

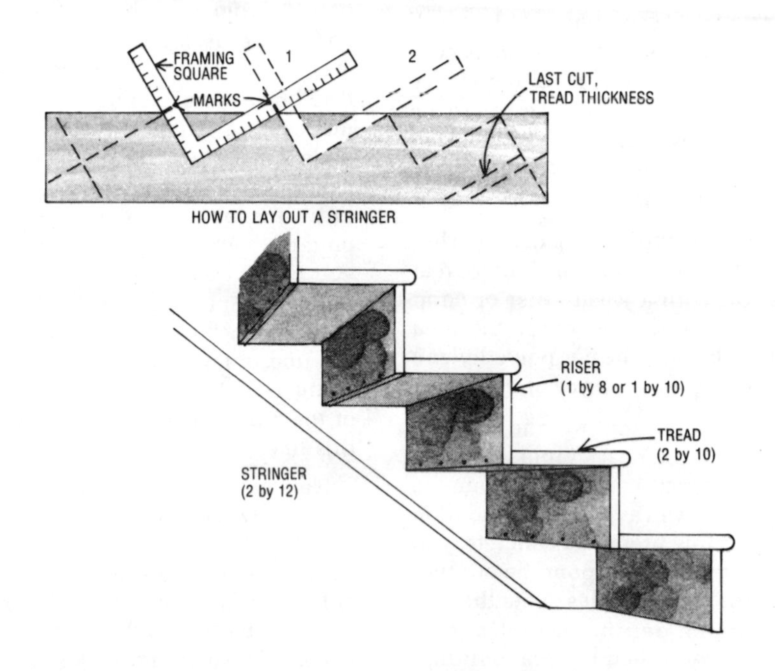

USE A FRAMING SQUARE to lay out the stringer for a simple stair.

How to Build a Deck

As an outdoor living area, a deck has a number of attractive advantages. For one thing, it is generally a cool location: a deck's frequent elevation above the ground allows for greater circulation of air. Then, too, decks are usually easier to sweep and keep clean than patios are. Finally, because of their linear and textural interest, decks have built-in style. The clean, flat lines of a deck and the warmth of natural wood are an asset to any home.

Some special techniques normally outside the skills of the carpenter must be considered during the initial phases of building a deck. These include preparing the deck site and constructing a sturdy concrete foundation.

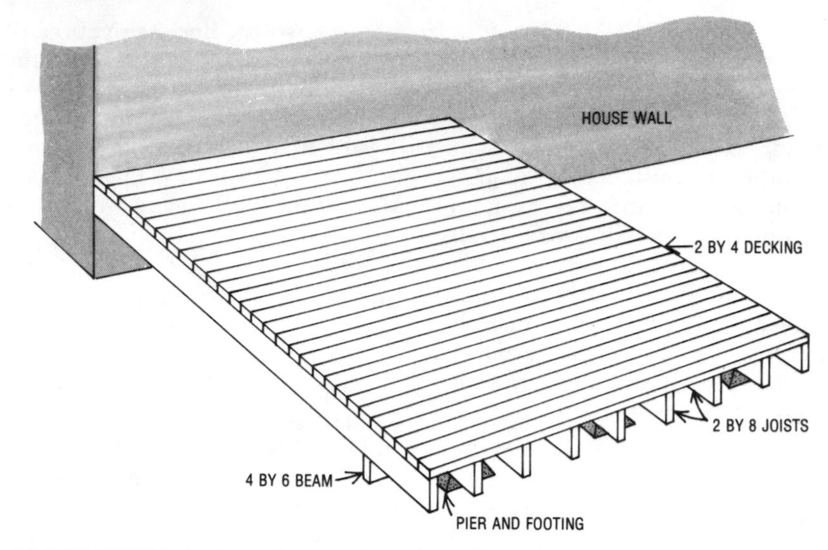

SIMPLE DECK is basic in design. This one is planned for level lot.

You can probably prepare the site yourself unless your yard presents large-scale grading or drainage problems. If you can't level the yard using a shovel and wheelbarrow, and if you don't feel that a couple of gravel-filled trenches leading out from under the deck will eliminate the possibility of standing water beneath the deck, it is probably best to consult a contractor.

The lot does not have to be level, but having it so simplifies deck construction. Discrepancies in distance from deck to foundation can be taken up by using differently sized lengths of 4 by 4 posts between the piers and beams.

Take one further step in preparing the site: weed control. Either treat the soil with a weed killer or chop out the weeds and cover the area with a layer of heavy polyethylene topped by a thin layer of gravel.

Setting the footing and foundation piers can be a simple job if you use premixed concrete to pour the footing as well as precast foundation piers. Using premixed concrete, you just add water and pour the mixture into the footing holes. Note that the minimum depth of footings is strictly controlled by local building codes. Check these codes, and, by all means, secure a building permit for the job.

Laying out the location of the deck is the first major construction step. This is usually done using stakes and string. Mark the proposed width and height of the deck on the house wall. If you'll have a door opening onto the deck, you'll probably want the deck's height to be level with the house's floor height. From a height mark at each end of the deck's proposed width, measure down the thickness of the decking material and make a second mark. Tie a string tightly between nails at these two second marks and, checking it with a level, adjust it until it is level. If a door will open onto the deck, check now to be sure you haven't adjusted the height so that it will place the deck's surface higher than the interior floor. Once the line is level, snap a chalk line along the string's span. The top edge of the 2 by 8 ledger will run along this line.

Next measure down the height of the joists (7¼ inches) plus the beams height (5½ inches) and make another set of marks, one at each end of the deck's proposed width. Drive a nail partially into the wall at each mark. Measure the deck's length (this one is 12 feet, 8 inches maximum) out across the yard from each nail and drive a tall stake (about 3-feet) into the ground at each spot. These stakes mark the two outer corners of the deck. To insure that the deck will stand square to the wall, use the 3-4-5 measuring method as shown.

When the stakes are square and at the proper distance from the house, tie a string tightly from each nail to its respective stake and level the string.

Note: If the decking is not spaced to allow water runoff, plan at this point to slope the deck slightly away from the house. The heights of the two strings tied at the stakes should also be level with each other if the first two leveling steps are done properly. Tie a string between the two stakes and check; if not level, adjust.

The strings mark the height of the top of the block on each pier where the beam will rest. Locate the position of the piers according to the plan, dig holes for the footings deep enough to satisfy codes, and pour the concrete. When the concrete is plastic enough to hold the precast piers where you place them, put the piers on the footings and adjust them to the proper height. Be sure all piers are level.

If necessary, you can use another string for checking pier heights once you have them into approximate position. Run it across all pier tops, not touching them but barely touching the undersides of the two strings extended from the wall, and tie it tightly between the two new stakes (shown in sketch).

When the concrete dries, you can begin building with the different wooden framing members. For a long-lasting deck, the members should be either decay-resistant redwood or standard framing lumber (like Douglas fir), pressure-treated with preservative. Follow the sketches for the rest of the wooden framing steps.

The deck described here is a simple, basic one. If you wish, you can vary the design, but be sure to stay within local building codes. For additional deck building ideas, consult the Sunset Book HOW TO BUILD DECKS.

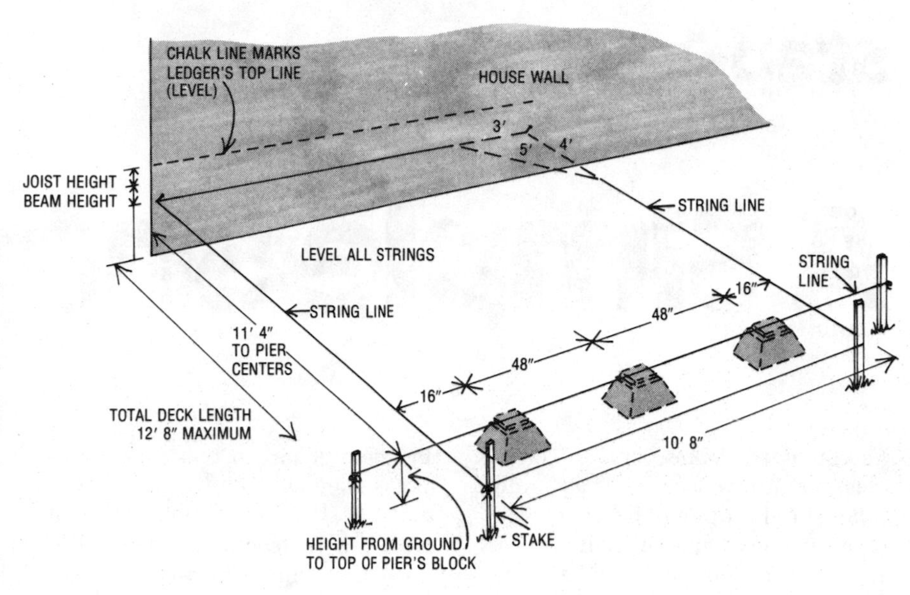

USE LEVEL STRINGS to lay out exact location of deck and foundation.

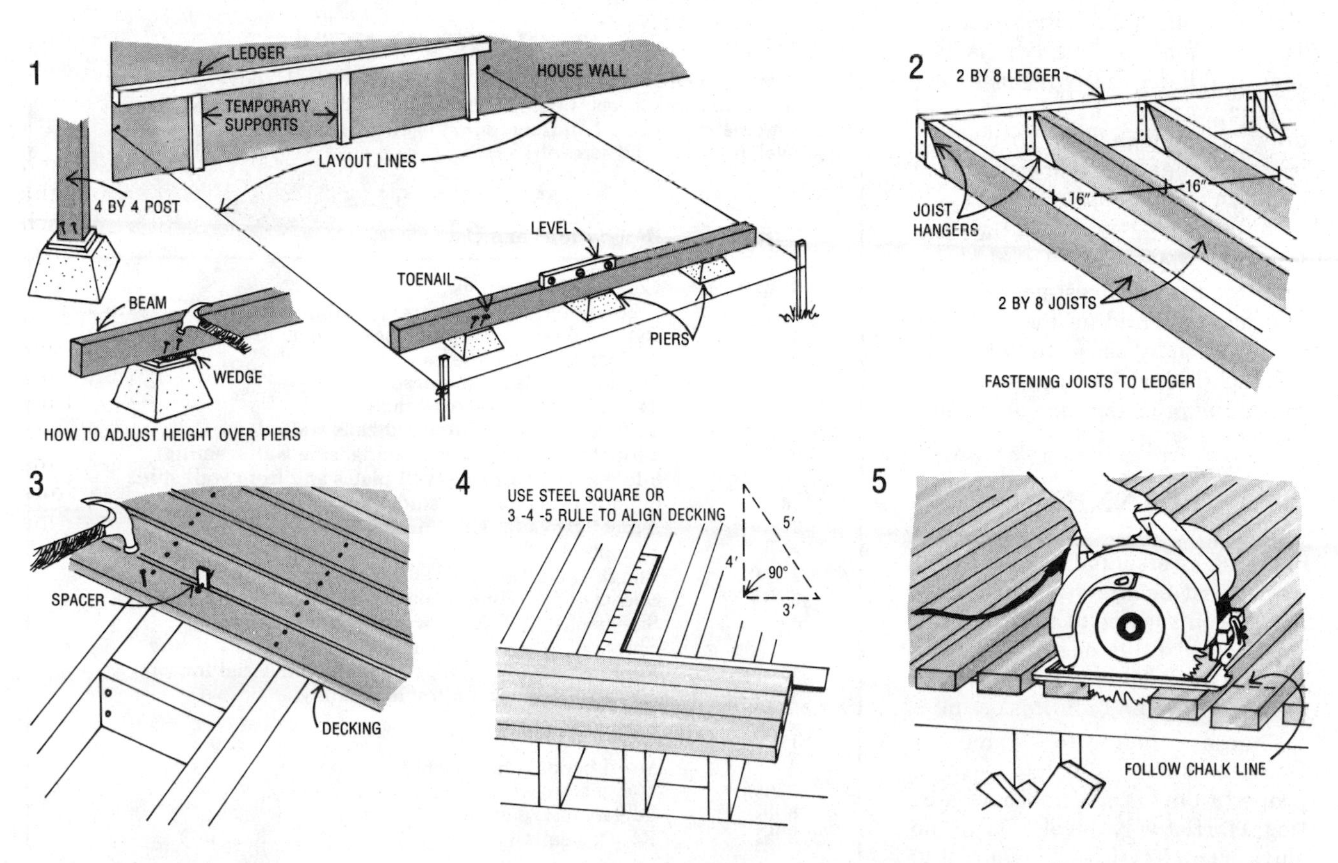

FASTEN LEDGER to wall and beam to piers (1). *Attach joists to ledger using joist hangers* (2). *Cross-brace joists across deck's center with extra 2 by 8's; make gauge for spacing decking from nail and yardstick* (3). *Check decking alignment using square or 3 -4 -5 - rule* (4). Trim edges along chalk line; seal ends (5).

Step-by-Step:

A Simple Shed

Lawnmowers, rakes, shovels, bicycles, plant pots, fertilizer bags, and other such equipment have a way of chopping away at your living space. A home of their own is what they need—and this shed is just the solution. Not only is it easy to build at a moderate cost but also it offers 8 by 12 feet of floor space—more than enough for organizing a mountain of clutter. And it's versatile. The broad front door admits just about anything that will fit inside; a long window along one wall draws in enough light to make it usable as a garden work shed or a children's playhouse.

Don't let the scale of the project put you off. Although it does require an investment of time and money, building the shed can be surprisingly easier than it looks if you follow instructions and tackle the processes one at a time.

PLANNING

Planning is essential to successful work. First check with your local building inspector to find out if a structure like this one is allowable in your area. Then ask whether or not you'll need a building permit.

Choose the site carefully. Don't locate the site too close to property lines, and choose an area that is relatively level. Place the shed where it will be convenient to use.

Think through the construction beforehand to make sure you have adequate tools and materials for the job. Front and rear walls are 12 feet wide; side walls are 8 feet wide. For water runoff, the roof slants from an 8-foot-high wall in front to a 6-foot wall in back. The door is 4 by 7 feet, the window 20 by 60 inches.

The shed was designed for easy construction at a relatively low cost. Rather than using a series of floor joists such as those typical floors have, this floor consists of

(Continued on page 82)

TOOLS YOU'LL NEED

Here are the main tools you'll need to build the shed: shovel, wheelbarrow, string and stakes, stepladder, tape measure, pencil, chalk line, combination square, handsaw (portable circular saw is very helpful), level, hammer, and screwdriver.

MATERIALS LIST

Quantity	Description* and Use
9 cu. ft.	Concrete for footings
6	Precast piers, 6 inches high, for foundation
1	Redwood 4 by 4 for posts (see page 82 to estimate length)
3	12-foot 4 by 6s for beams
12	10-foot 2 by 6s for roof framing
4	14-foot 2 by 4s for crossbracing
6	12-foot 2 by 4s for front and back wall plates
6	10-foot 2 by 4s for crossbracing, side wall framing
18	8-foot 2 by 4s for side wall plates and front wall studs
18	6-foot 2 by 4s for wall studs
3	5-foot 2 by 4s for sills and headers
2	12-foot 2 by 2s for door framing
210 ft.	Rough-sawn 1 by 3 for battens
3	Sheets of 1⅛" Tongue-and-groove plywood for flooring
4	Sheets of ½" C-D-X plywood for roof sheathing
10	Sheets of ⅜" resawn Douglas fir plywood for siding
1½ sqs.	Roofing felt (90 lb.) or other roofing material for roof
1	20" by 60" aluminum-frame window
3	Door hinges
1	Latch
2	Metal bands (see page 86)
12 lbs.	16d galvanized box nails
8 lbs.	8d galvanized box nails
2 lbs.	Roofing nails

*For all framing, use "Standard" Douglas fir or "Common" Southern Pine.

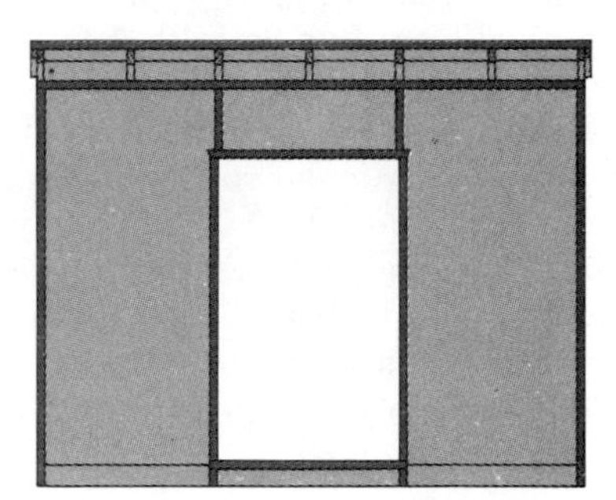

FRONT VIEW

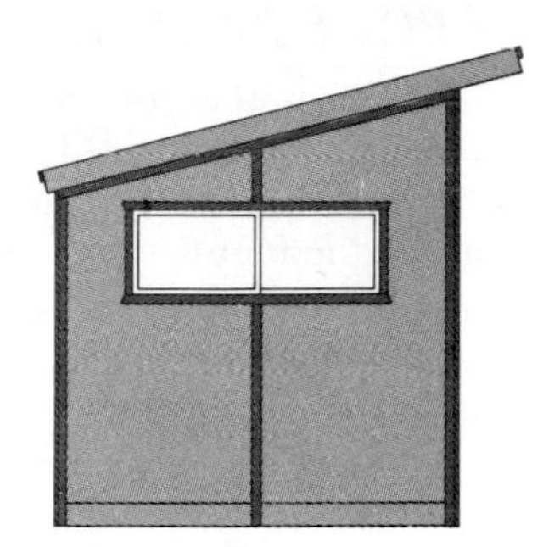

SIDE VIEW

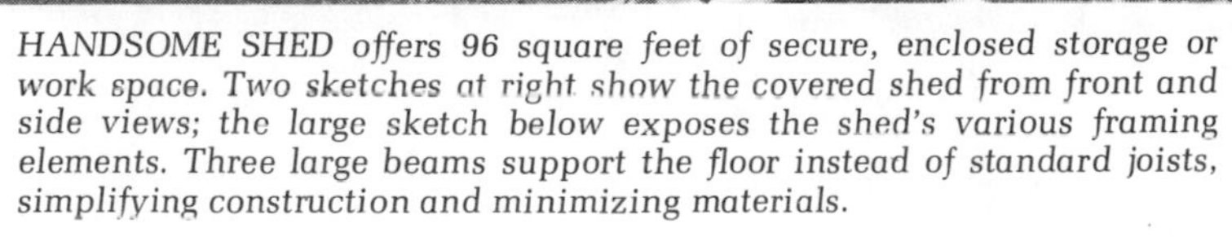

HANDSOME SHED offers 96 square feet of secure, enclosed storage or work space. Two sketches at right show the covered shed from front and side views; the large sketch below exposes the shed's various framing elements. Three large beams support the floor instead of standard joists, simplifying construction and minimizing materials.

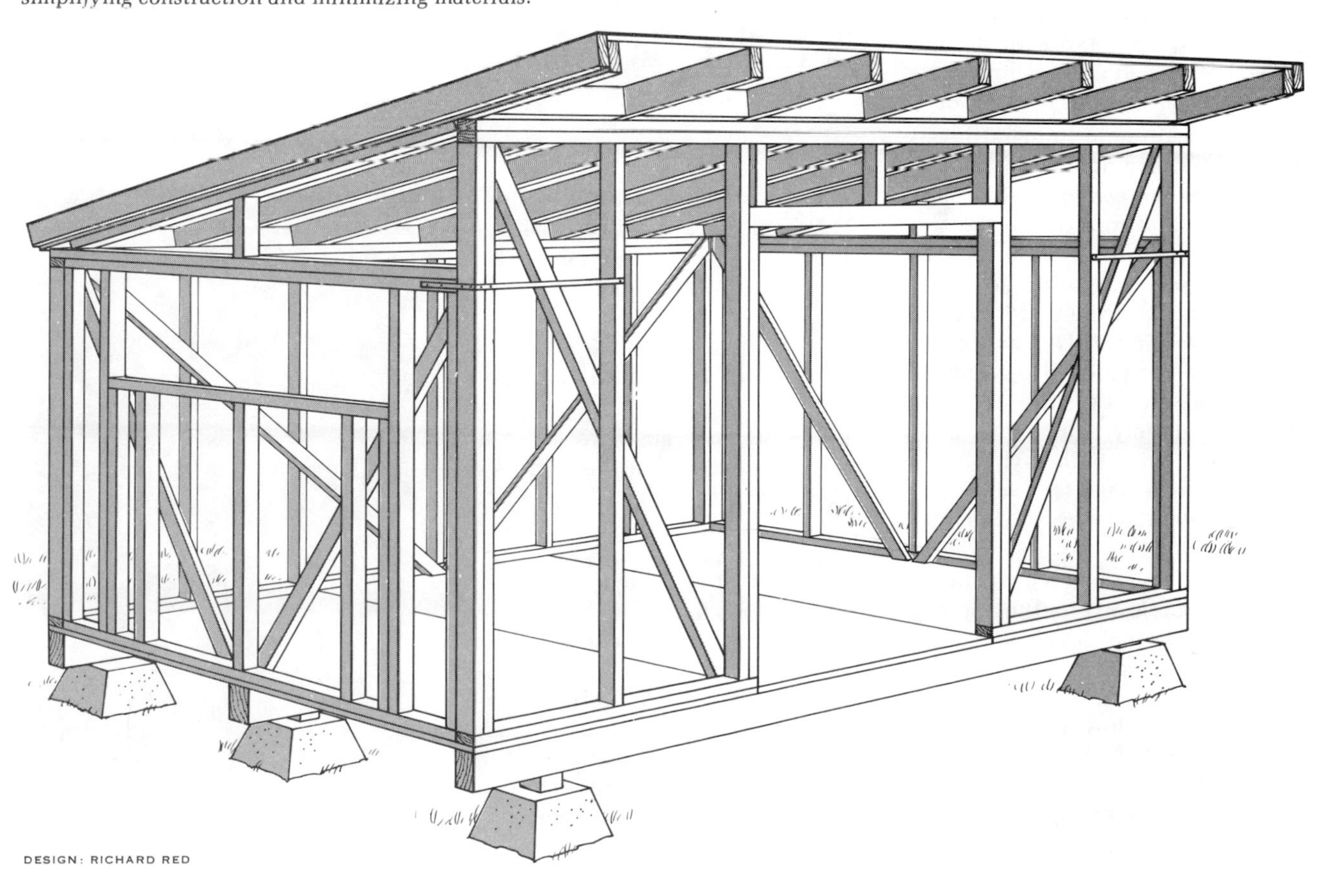

DESIGN: RICHARD RED

three 4 by 6 beams, laid across concrete piers and then covered by 1⅛-inch tongue-and-groove plywood. The walls are framed with 2 by 4s on 24-inch centers, and the roof framing is of 2 by 6s on 24-inch centers. This particular roof is covered first with ½-inch C-D grade plywood and then with a layer of composition shingles. The exterior walls were faced with rough-sawn Douglas fir plywood, easy to apply and rustic looking.

LAYING THE FOUNDATION

Use methods similar to those discussed under "How to Build a Simple Deck" on page 78 to prepare the site and lay the foundation.

Clear the area and level off large bumps or holes. To represent the four walls of the shed, stretch strings between stakes, placing the stakes about 1 foot beyond actual widths of the walls (see sketch at right). Check for square using the 3′-4′-5′ method shown in step 4 on page 79. (Another way to make sure the strings are at 90° angles to each other is to match diagonal measurements, as shown in the photograph under "Framing the Floor.")

Break the ground to mark the six footings according to the illustration. Place the center of each footing and pier 1 foot inside the side wall strings (the 4 by 6 joists span only 10 feet, according to code). The side walls of the building overhang the footings this distance. Dig the holes 8 inches deep and 18 inches square. Be sure to place them accurately. All piers don't need to be level with each other—you later make up for height discrepancies by adding short 4 by 4 blocks.

To keep the shed closer to the ground, you can dig the footings deeper, lowering the piers into the holes. (This shed's piers are 6 inches high; the footing holes are 12 inches deep, allowing 8 inches for concrete and 4 inches for piers. This leaves only 2 inches of each pier showing above ground level.)

Mix the concrete and pour it 8

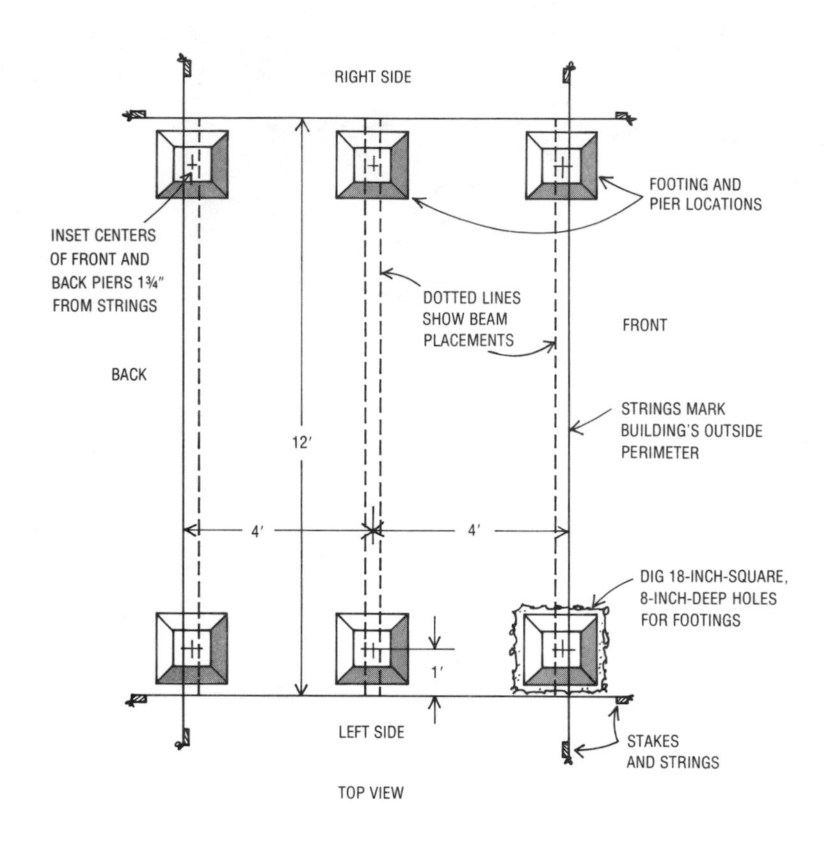

STAKES AND STRINGS guide placement of footings and piers. Notice front and back piers are inset 1¾ inches; side piers are inset 1 foot.

inches deep into the holes. Allow it to become firm yet tacky; then carefully place the piers. Level their tops and allow the footings to dry.

FRAMING THE FLOOR

When the concrete dries, make up for discrepancies in the heights of piers by adding short lengths of 4 by 4 redwood posts to the tops of the low piers, bringing each up to the level of the highest one. Arrange for the length of each post by placing one end of a long, straight 2 by 4 on the highest pier and then checking it for level as you hold the other end over each pier top. Measure the distance from the board's underside to the pier's top (see photo on opposite page).

Place the short posts on their

SET PIERS in concrete when it has enough body to hold them. Let concrete dry thoroughly before continuing.

MEASURE for post lengths from pier tops to underside of a board that is level with the top of the highest pier.

piers but don't nail them. Set the 12-foot, 4 by 6 beams onto piers and posts and check each individual beam for level. Then span the long, straight 2 by 4 from one beam to

another to see if they are level with each other. If they are not, cut new posts or later adjust heights by wedging shingles or small blocks between post tops and beams.

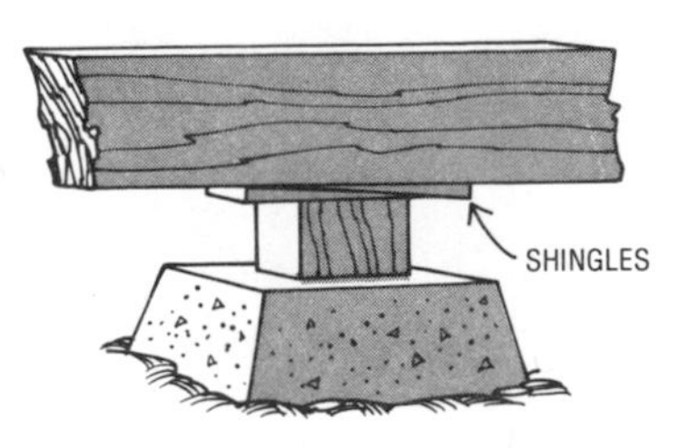

Make sure each beam is centered exactly over the piers. Then check for square by measuring diagonally from opposing corners as shown in the photograph below. Measurements should match.

Toenail posts to beams and to redwood pads on pier tops (three 16d galvanized nails at top and bottom), making sure you don't move the beams. A pencil mark from beam to post to pier pad will help keep everything lined up. Once the beams are nailed, you can remove the guideline strings and stakes.

COVERING THE FLOOR

Lay the three 4 by 8-foot, 1⅛-inch-thick plywood sheets in position on the beams, good side up and lengthwise across all three beams. The tongue of the first panel should face center. You may have to nail the first sheet into place before you can fit the others (use 16d galvanized nails every 8 inches). Snap a chalkline over the beam so you know where to nail.

TONGUE of first panel faces inward.

If the second sheet's tongue doesn't fit easily into the groove, place a 4-foot scrap 2 by 4 against it and hammer it into place. When all three sheets are placed, cut off the tongue of the third sheet. Nail them all down.

LAYING OUT WALLS

This structure's walls are assembled flat on the floor and then raised into position. This makes marking and nailing them easy work.

First you'll need to mark the edges of the sole plates and top plates for the placement of wall studs. Lay pairs of 2 by 4s flat, flush with the floor's edges as shown in the sketch. Temporarily tack them

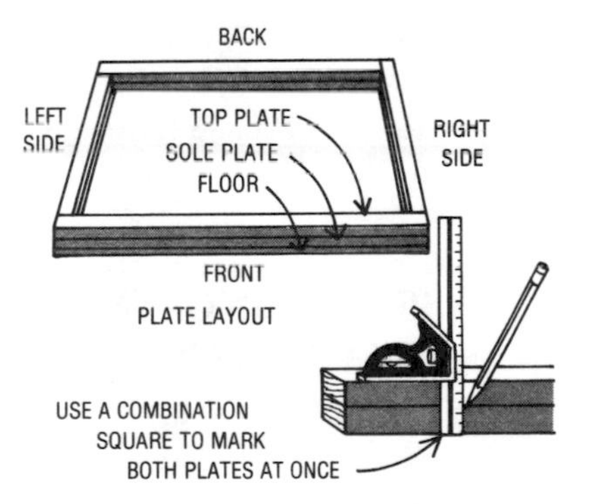

down with two or three 8d nails so they won't move. Notice which plates butt into others and cut off the ends of some for proper fit. These plates should cover the floor's perimeter.

Next, for marking locations of wall studs, window opening, and door opening, you'll need a tape measure, pencil, and combination square. Mark both plates simultaneously along one edge as shown, using a combination square. Duplicate the marks in the sole plate sketch on the following page

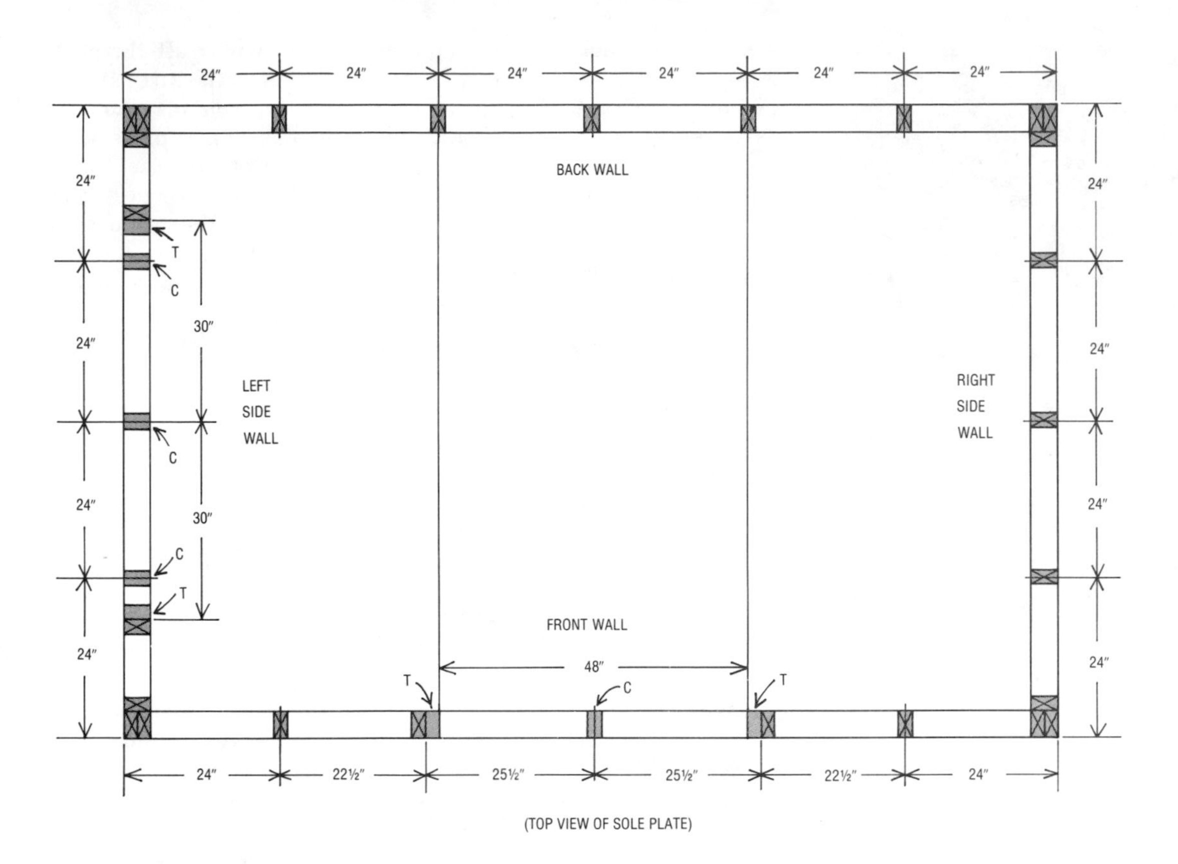

MARK WALL STUD PLACEMENTS on sole plates and top plates where indicated. The "Xs" show where to put full-length studs, "Ts" are for trimmer studs, and "Cs" are for cripple studs. Right side and back walls have studs on regular 24-inch centers.

—"Xs" mark placement of full-length studs, "Ts" mark for trimmer studs, and "Cs" mark cripple studs.

ASSEMBLING AND ERECTING THE WALLS

Assemble the walls one at a time, beginning with the front wall. To do this, remove all of the 2 by 4 plates but the pair you'll be working on. Separate those two plates; one will be the top plate, the other will be the sole plate. Lay them on their edges at opposite ends of the floor and place the various full-length studs between them. (Illustrated on the facing page are the stud placements of the front and left side walls; these both contain cripple and trimmer studs. Since the right side wall and rear wall contain only full-length studs, they are not illustrated.) First you will have to cut a few inches off most of the studs.

Nail two 16d galvanized nails through plates into ends of each stud.

Using 16d nails, nail through the plates into the ends of the studs, keeping studs placed exactly on their marks. Also nail all of the other wall framing members except the trimmer studs. On the front wall, nail the cap plate onto the top plate before raising it into position; on the other walls, wait until after raising them to add cap plates.

BRACE, temporarily nailed to end stud, helps support wall as you stand it up.

Have a friend help you raise the wall and hold it while you tack temporary braces to both ends to hold it in position (see photo).

(Continued on page 86)

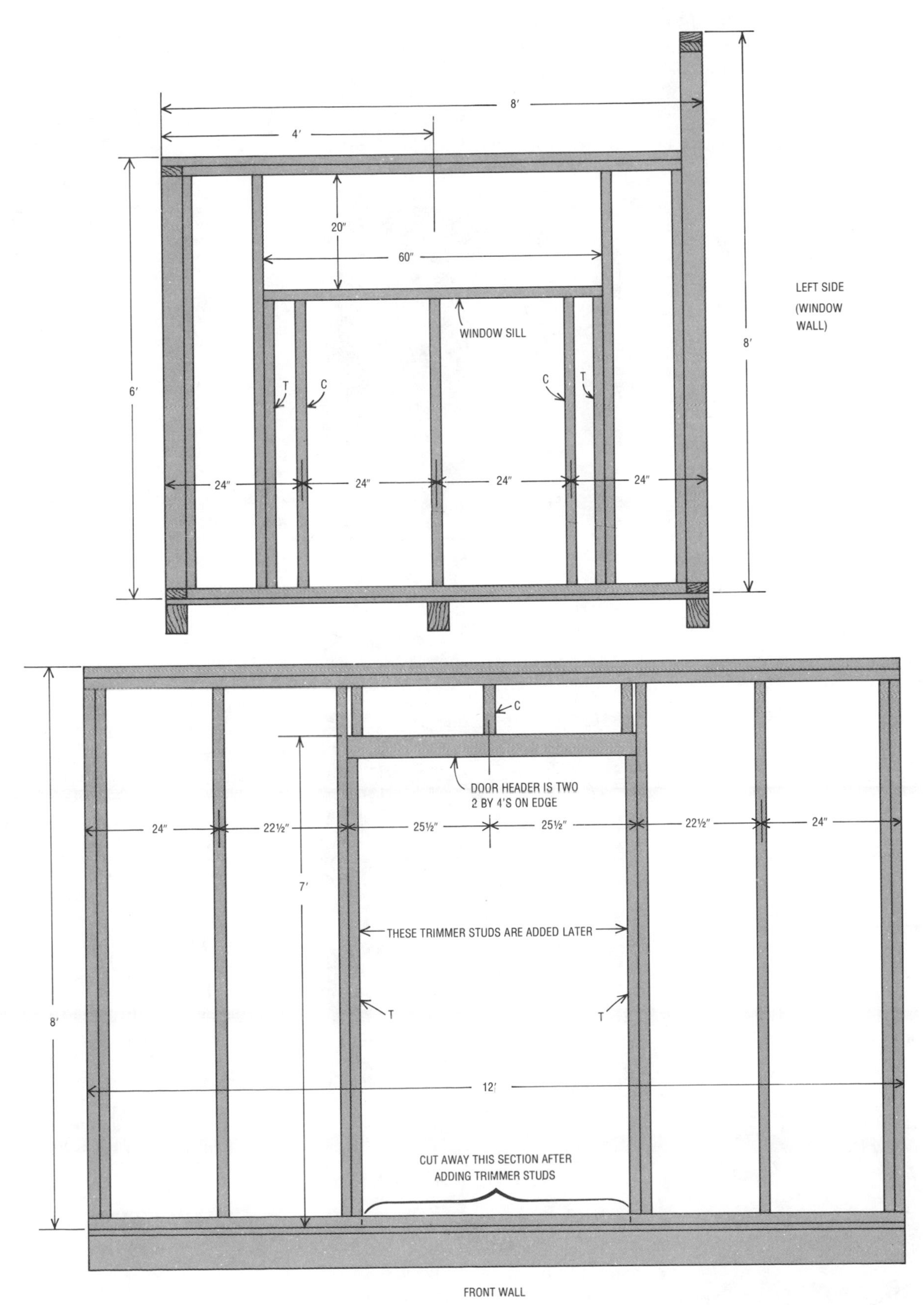

STUD LAYOUTS for front and side walls are illustrated above. Assemble the walls flat on the floor and then raise them into position. Don't add trimmer studs or headers until walls are erected; delay attaching cap plates of side and back walls.

Align the sole plate perfectly flush with the outside edge of the floor (see sketch); then nail it to the

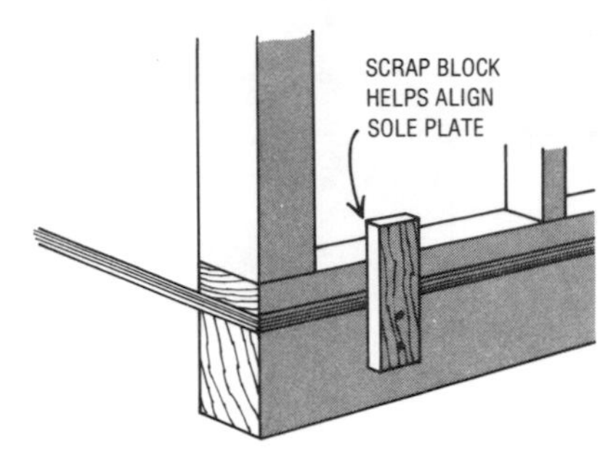

floor with two 16d nails between each pair of studs (at the door opening, put two 16d nails within 1½″ of the studs—you'll be removing the rest of the plate there).

After you finish the front wall, remember that the other three walls are only 6 feet high.

Assemble and raise the right side wall, nail it to the floor as before, and face nail it to the adjoining stud of the front wall (six 16d nails).

Then assemble and raise the back wall and left side wall (in that order). When all walls are up, add the remaining cap plates, overlapping

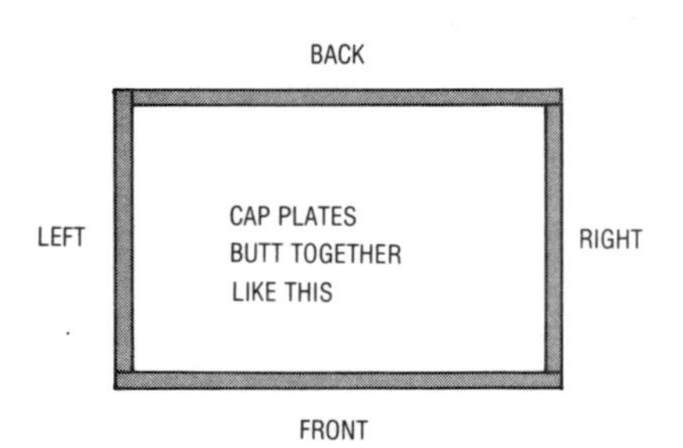

them as shown. Because you can't interlock the front wall, secure it to the two side walls with a metal band (some of the material for building the shed may come wrapped in such a band—if not, ask for a couple at the lumberyard).

Use the level to get the walls perfectly plumb; then brace them

with diagonal crossbraces inside as shown. You can then remove the temporary supports.

PLACING RAFTERS

The roof framing consists of seven 10-foot, 2 by 6 rafters that run from front to back, one additional 2 by 6 facia on each side, and two rows of frieze blocks. The primary rafters and frieze blocking are illustrated.

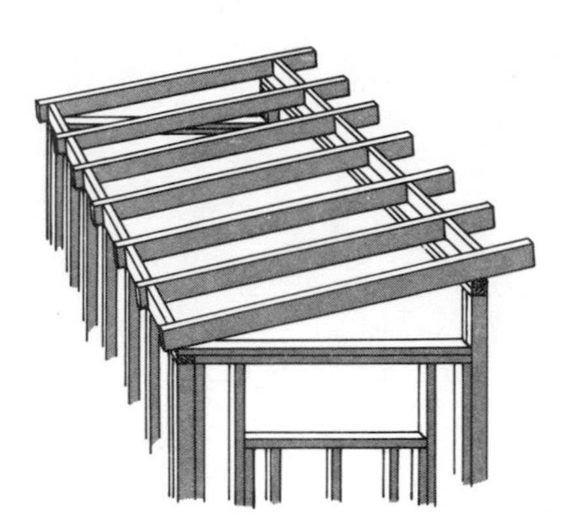

Stand on a stepladder to mark positions for the rafters on the cap plates. Just use a combination square to transfer the regular 24-inch-apart stud marks for the front and rear walls to the cap plates (the rafters will run directly over the studs).

Once all seven rafters are exactly 10 feet long, make a mark 3 inches from one end of each. This is the distance each will extend beyond the rear wall.

Fasten the outside rafters first. Place them flush to the outside of the side walls and toenail them at each end, using two 16d nails on one side and one 16d nail on the other side (at both ends).

No need to notch the rafters where they contact the cap plates. Their short span, the light weight of the roof, and the use of frieze blocks eliminates that need.

The frieze blocks are short lengths of 2 by 6 placed between rafters to keep them from twisting or turning and to seal the space between the roof and the cap plates. Cut them to fit exactly and nail them flush to the outside of the cap plate and between the rafters.

COMPLETING THE FRAMING

Once wall and roof framing are up, additional framing must be completed before you can enclose the roof and walls of the structure. First, face nail a 2 by 4 to the lower edge of the two end rafters (see sketch). Then place a 2 by 4 cripple stud between each of those 2 by 4s and the cap plate of each side wall, four feet on center from each end of the wall. Cut one end of the cripple stud at a 14° angle so it will fit against the slanted 2 by 4 above it. Toenail it into place.

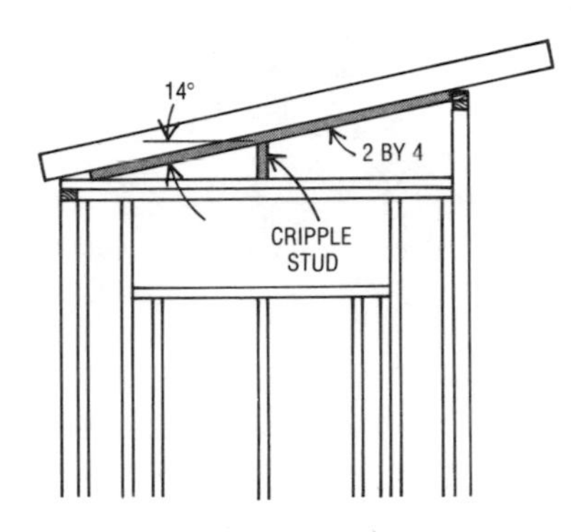

Nail the two remaining 2 by 6s to the outside rafters, allowing these facia boards to protrude ½-inch above the top of the rafter

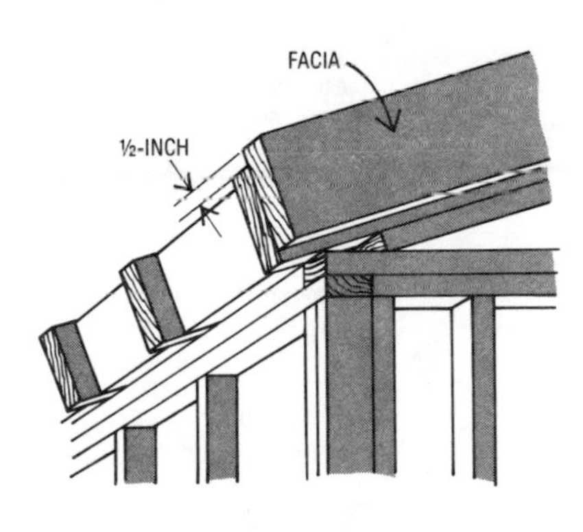

(this extra will butt against the ½-inch plywood roofing). Use 16d galvanized nails.

Cut the two door-trimmer studs to fit snugly beneath the door header. Toenail the trimmers at the top and bottom—this keeps them from twisting—and face nail them to their neighbors. Then cut and remove the section of sole plate in the doorway.

Install the window as explained on page 68. Although this window is lower than normal, its placement provides the simplest solution to framing.

Here's a helpful tip for nailing the aluminum-frame window: first nail the corners by driving 8d nails through the flange. Then slide it about ¼-inch open. If there are any broad gaps, wedge the window until all edges are parallel and nail the flange near the center of the window, top and bottom.

This ends the framing processes. Check all framing for unnailed areas, improper fitting, and crooked framing members. Then proceed to enclose the shed.

COVERING THE ROOF

Lay the ½-inch plywood sheets on the roof framing in position as shown. You can stagger the joints

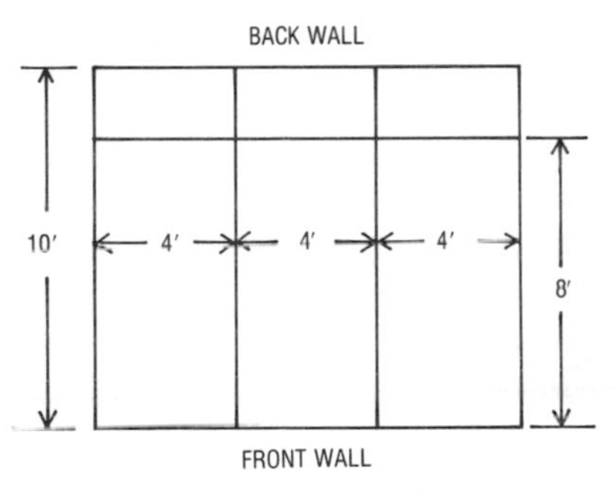

PLYWOOD LAYOUT, ROOF

of the plywood as illustrated on page 43 and 73 for added strength. Snap a chalkline above the rafters to guide your nailing and nail with 8d nails every 8 inches.

Place 1 by 3 redwood battens along the front and rear edges of the rafters, flush to the top of the sheathing. This will cover the exposed plywood edges and dress-up the rafter ends. Nail the 1 by 3s with 8d galvanized nails.

The roof covering is up to you. You can shingle the roof as explained on page 45, apply composition shingles, or use 90-pound roofing felt. The 90-pound felt is less expensive, easy to apply, but not as durable (it will last about 3 years in most areas).

COVERING EXTERIOR WALLS

When cutting and nailing the siding, work carefully. Mistakes are more evident here than in any other part of the building. Remember, when cutting the siding to match the slope of the roof, don't cut the angles backwards. Only one side of the panels can face outward. (See page 26 on hints for working with plywood.)

Nail the siding to the studs using 8d nails. If you wish, you can hide the beams from view by first nailing a skirt of siding around the base of the structure. When all

BATTENS from 1 by 3s cover joints where plywood siding panels meet. For their placements, see page 81.

of the plywood is nailed in place, add the 1 by 2 battens over the seams and nails.

Make the door as shown in the sketch, allowing about ⅜-inch clearance on all sides.

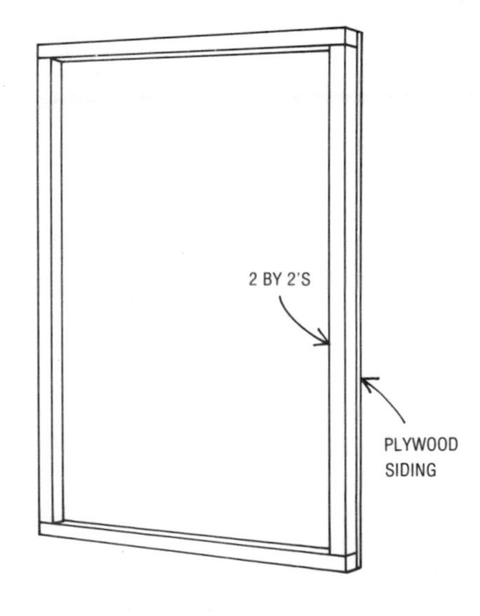

INDEX

PHOTOGRAPHERS: **Ernest Braun:** 14, 15, 16 top left, 20 top, 21 center left and right, 22 top left and right, 23 top right. **Glenn M. Christiansen:** 8, 11 top right. **Ells Marugg:** 21 bottom right. **V. Lee Oertle:** 12 top left. **Norman A. Plate:** 11 top left. **Richard Red:** 81, 82, 83, 84, 86, 87. **Donald W. Vandervort:** 6, 7, 9, 10, 12 top left, 13, 20 bottom, 21 top, bottom left, 22 bottom, 23 top left, bottom. **Darrow M. Watt:** 16 top right, center bottom, 17, 18, 29, 30, 31. **Western Wood Products Association:** 27, 28.